Gerhard Gierenz, Werner Karmann
(Editors)

Adhesives and Adhesive Tapes

WILEY-VCH

Other Titles of Interest:

R. Mildenberg, M. Zander, G. Collin
Hydrocarbon Resins
First Edition 1997
ISBN: 3-527-28617-9

Wässrige Polymerdispersionen
Edited by D. Distler
First Edition 1999
ISBN: 3-527-29587-9

In Preparation:
Applications of Emulsion Polymers
Edited by Dieter Urban
and Koichi Takamura
First Edition 2001
ISBN: 3-527-30312-X

Gerhard Gierenz, Werner Karmann (Editors)

Adhesives and Adhesive Tapes

WILEY-VCH

Weinheim · New York · Chichester · Brisbane · Singapore · Toronto

Dr. Gerhard Gierenz
Dieselstraße 80
D-42719 Solingen
Federal Republic of Germany

Dr. Werner Karmann
Beiersdorf AG
Unnastraße 48
D-20245 Hamburg
Federal Republic of Germany

Library of Congress Card No.: applied for

British Library Cataloguing-in-Publication Data:
A catalogue record for this book is available from the British Library

Die Deutsche Bibliothek – CIP-Einheitsaufnahme
A catalogue record for this book is available from Die Deutsche Bibliothek

ISBN 3-527-30110-0

Printed on acid-free paper.

Composition: Rombach GmbH, D-79115 Freiburg
Printing: strauss-offsetdruck, D-69509 Mörlenbach
Bookbinding: Wilhelm Osswald + Co., D-67433 Neustadt

Preface

The impressive developments in numerous fields of technology – anything we can imagine from bookbinding, food packaging, electronic devices, automobile manufacture to building and aircraft constructions – all engeneering progress in these and other areas is enabled and accelerated by innovations in adhesive technology.

Although adhesive applications mostly are not readily evident in daily life – except for some do-it-yourself areas, e. g. paper gluing, and repairs – experts in the development and engeneering departments value adhesive technology as an important tool to find solutions to a wide range of technical problems. Strong points are the great variability in the kind and the properties of the adhesives, e. g. drying or reacting liquids, melting solids, adhesive tapes and die cuts.

This book, a special edition of the chapters "Adhesives" and "Adhesive Tapes" of Ullmann's Encyclopedia of Industrial Chemistry, is intended as a guidance for everybody interested in getting an overview on the various adhesives, their different properties, and the wide field of applications.

Gerhard Gierenz　　　　Werner Karmann

Acknowledgement:

We would like to thank for their valuable contributions to Dieter Dausmann, Wolfgang Dierichs, Werner Gruber, Werner Haller, Ansgar van Halteren, Günter Henke, Michael Hirthammer, Horst Hoffmann, Matthew Holloway, Andreas Kummer, Christoph Matz, David Melody, Hermann Onusseit, Riza-Nur Özelli, Richard D. Rich, Ernst-Ulrich Rust, Hans-Peter Sattler, Günter Tauber, Lothar Thiele, and Udo Windhövel. Our particular thanks go to our companies tesa Beiersdorf AG und Henkel KGaA for sponsering and supporting our project.

Gerhard Gierenz   Werner Karmann

Contents

1. Adhesives

WERNER HALLER, Henkel KGaA, Düsseldorf, Federal Republic of Germany (Chaps. 1.1–1.3, Sections 1.9.7 and 1.9.10)

HERMANN ONUSSEIT, Haan, Federal Republic of Germany (Chap. 1.4, Sections 1.9.1, 1.9.2, 1.9.3, 1.9.4)

GERHARD GIERENZ, Solingen, Federal Republic of Germany (Chap. 1.5)

WERNER GRUBER, Korschenbroich, Federal Republic of Germany (Chap. 1.6, Section 1.9.9)

RICHARD D. RICH, Loctite Corporation, Rocky Hill, CT 06067, United States (Chaps. 1.6 and 1.8 in part, Section 1.9.21)

GÜNTER HENKE, Henkel KGaA, Düsseldorf, Federal Republic of Germany (Chap. 1.7)

LOTHAR THIELE, Langenfeld, Federal Republic of Germany (Chap. 1.8)

HORST HOFFMANN, Dorus Klebetechnik GmbH & Co. KG, Bopfingen, Federal Republic of Germany (Section 1.9.5)

DIETER DAUSMANN, Münchweiler, Federal Republic of Germany (Section 1.9.6)

RIZA-NUR ÖZELLI, Düsseldorf, Federal Republic of Germany (Chap. 1.9.8)

UDO WINDHÖVEL, Monheim, Federal Republic of Germany (Section 1.9.11)

HANS-PETER SATTLER, Kamen, Federal Republic of Germany (Section 1.9.11 in part, Section 1.9.12)

WOLFGANG DIERICHS, Düsseldorf, Federal Republic of Germany (Sections 1.9.13, and 1.9.15)

GÜNTER TAUBER, Düsseldorf, Federal Republic of Germany (Section 1.9.14)

MICHAEL HIRTHAMMER, Henkel Teroson, Heidelberg, Federal Republic of Germany (Section 1.9.16)

CHRISTOPH MATZ, Lufthansa Technik AG, Hamburg, Federal Republic of Germany (Section 1.9.17)

MATTHEW HOLLOWAY, Loctite RD & E, Dublin, Ireland (Section 1.9.18)

DAVID MELODY, Dublin, Ireland (Section 1.9.19)

ERNST-ULRICH RUST, Bergisch Gladbach, Federal Republic of Germany (Section 1.9.20)

ANSGAR VAN HALTEREN, Industrieverband Klebstoffe e.V., Düsseldorf, Federal Republic of Germany (Chap. 1.10)

1.1. Introduction

Adhesives are defined as nonmetallic substances capable of joining materials by surface bonding (adhesion), the bond itself possessing adequate internal strength (cohesion). Adhesive is a generic term and covers other common terms, such as glue, paste, gums, adhesive cement, and bonding agent.

History. Adhesion is among the oldest technologies of mankind [129]. As early as the Stone Age, our ancestors made stone axes and other tools using mineral pitch or wood rosin. Asphalt was used in building the Tower of Babel. In ancient Egypt, it was already customary to veneer furniture with the aid of animal glues. Flour- and casein-based adhesives were applied in ancient Rome. Adhesives of plant origin such as wood rosin were as well-known in China, as was the gum arabic and the caoutchouc ("weeping wood") from the tropical regions of South America and Asia.

In the Middle Ages, the first glue-boiling plants came into being and produced protein glues from animal raw materials (glutin from hides and bones, blood albumin, casein from milk) or starch paste from plants.

With the developments in chemistry in the early 1900s, the technology of adhesives began to boom. Phenolic resins (ca. 1900), melamine resins, and urea resins (ca. 1930), polymer dispersions, epoxy resins (1938), and cyanoacrylates (1957) largely superseded the classic natural adhesives and, with a multitude of bindings agents, form the basis for modern adhesive technology, which is one of the most advanced joining processes.

Composition. An adhesive is composed of basic raw materials, which are called binders [1] and which determine its adhesiveness (adhesion) and its internal strength (cohesion), and of frequently necessary auxiliaries, which establish particular end-use and processing characteristics. The adhesiveness of an adhesive, its internal strength after setting, and its processing characteristics are the fundamental properties that determine its suitability for use in forming adhesive joints. Adhesive joints are the joints formed between substrates and adherents using adhesives.

The binders used for adhesives are primarily high polymers having optimal strength properties. High internal strength (cohesion) is essential if the adhesive in an adhesive joint is to be able to transmit forces from one adherent to the other. Most adhesives contain high molecular mass organic substances as their basic raw materials or reactive organic compounds that are preliminary stages of polymers and that react during the bonding process to form polymers. Inorganic polymers, such as the various types of waterglass, are used only to a very limited extent.

Virtually any standard polycondensate, homopolymer, and copolymer and also polyadducts may be used, provided they can be applied as solutions, dispersions, emulsions, or melts. In addition to these raw materials, auxiliaries such as resins, plasticizers, fillers, thickeners, solvents, antiagers, preservatives, hardeners, or setting retarders, are required, depending on the end use. Their function is inter alia to adjust tack, to improve adhesion, to make flexible, to regulate viscosity, to stabilize, and to influence setting or hardening.

Adhesive Joints. In many cases, adhesion is as effective as other joining techniques, such as riveting, welding, soldering, and screwing; in some cases it complements those techniques; and frequently it affords numerous advantages. On the other hand, certain processing requirements and material properties related to chemical structure can restrict the use of adhesives.

Bonding has the advantage that the joining of adherents can be carried out very rationally, often extremely quickly, and particularly economically [2]. One important feature common to all adhesive joints is the highly uniform distribution of forces over the entire joint area by comparison with rivet and screw joints. The uniform stress levels frequently provide for more optimal utilization of material strength. Accordingly, thinner and hence lighter sections of material may be used. This has resulted in the saving of material and, for example, in the automotive, aircraft, manufacturing, and furniture industries, in totally new structural elements (sandwich elements).

The use of adhesives enables numerous materials, even those differing widely in type, to be joined to one another. This applies above all to materials that cannot be joined to one another by other techniques. Because some adhesives set even in relatively thick layers, any unevenness in the constituent material of the adherents may be smoothed out during bonding, or significant dimensional tolerances may even be bridged by the gap-filling principle. The adhesive layer of an adhesive joint even may have a vibration-dampening effect and, by virtue of its insulating properties, can prevent contact corrosion in adhesive joints involving metals of different normal potential. Another advantage is the fact that many adhesives can be applied at room temperature or, when heat must be used for application, it is sufficient to apply temperatures at which the constituent materials of the adherents are not affected, as can happen in the welding of metals and plastics.

The usability of adhesives can be restricted when stringent requirements are imposed on the thermal stability of an adhesive joint [2]. Like all plastics, adhesives based on organic polymers also show a marked dependence on temperature in their strength properties. The strength properties of an adhesive joint can be kept constant and sufficiently high in only a relatively narrow temperature range. Adhesives cured by thermal cross-linking which are still relatively temperature resistant are generally capable of withstanding temperatures of up to about 150 °C. Using new, but more difficult to apply adhesives based on polyimides [3], it is possible to form adhesive joints which are capable of withstanding temperatures of the order of 250 °C for long periods. On the other hand, some adhesives, when subjected to heavy static stress for long periods, show a tendency even at room temperature to creep, a plastic deformation that can result in separation of the adhesive joint.

1.2. Adhesion Theories

The atoms or molecules in any solid are held together by so-called cohesion forces. These interatomic and intermolecular forces are conservative in character, i.e., not dependent upon the pretreatment of the material [4, p. 1]. After a break in the solid, the two fragments cannot be put back together in such a way as to restore the solid to its original state. Reactivation of the cohesion forces would be possible only if the original intervals between atoms or molecules could be reestablished. The same may be said also of the joining of two different materials; a mutual convergence over the entire joint area on the radius of interaction of the molecules would result in adhesion of the same order of magnitude as the cohesion forces [5]. In practice such convergence is not possible. There are three reasons for this: first, the surface roughness of the adherents; second, a reduction in the effective surface for boundary layer reactions due to the surface tension of the adhesive; and third, the formation of the so-called weak boundary layer [130], [131]. This is a region within the adhesive film close to the boundary layer where the reactive groups of the molecules are directed to the adherent and consequently do not contribute to the curing process, that is, no cross-linking occurs. Therefore, the weak boundary layer is always a weak link in the adhesive joint. The function of the adhesive in the boundary zone is to compensate for the effects of surface roughness by providing optimal wetting of the adherents.

Over the years, several different theories have been developed as to the mode of action of adhesives, i.e., the mechanism of adhesion. They extend from the simple mechanical theory via the electrostatic and adsorption theory to the diffusion theory, these being the most important [6]. One feature common to all these theories is that, for some adhesion phenomena, these theories allow qualitatively satisfactory assertions to be made, but in other respects they fail totally and, in some cases, even lead to conflicting results. The adhesion of adhesive to an adherent, on which any joint is based, is clearly not a consistent and isolatable process [7] but a complex addition of various adhesion effects. Because the number of individual processes involved, apart from exceptions, is extremely difficult to estimate, definitive confirmation or rejection of individual theories is hardly possible.

Mechanical Theory of Adhesion. The oldest theory of adhesion is definitely the mechanical theory. It is based on mechanical anchorage of the adhesive in pores and irregularities in the adherent and is discussed primarily in reference to wood and similar porous materials [8].

Electrostatic Theory of Adhesion. According to the electrostatic theory [9], the adhesion forces between adherent and adhesive layer are applied by contact or transfer potentials. These transfer potentials cause the buildup of an electric double layer at the adhesive – adherent boundary and corresponding Coulomb attraction forces between the two components. In principle, the occurrence of transfer potentials is unquestionable, as shown by the electrostatic discharges that can be detected in the destruction of adhesive joints. However, the practical significance of the attraction forces associated with these discharges is still being debated [4, pp. 150 – 153], [10].

Adsorption Theory. As its name indicates, this theory of adhesion draws upon surface forces for explaining the observed phenomena [11]. It regards adhesion as essentially a special property of phase interfaces. The forces that are responsible for adhesion in this process are the so-called secondary valence or van der Waal's forces. These forces have three components, namely Keesom's dipole orienting effect, Debye's induced dipole effect, and London's dispersion effect [12]–[14]. For these forces to become active, the distances between the molecules of adhesive and adherent must converge toward molecular intervals. This requires complete spreading of the adhesive over the surface of the adherent. Good adhesion can be expected if the adhesive (in liquid form or in a liquid medium) wets the adherent [8]. Accordingly, the adsorption theory may be placed on an entirely thermodynamic basis. Adhesion is thus determined by the ratio between the surface energies of the adhesive and the constituent material of the adherent in the sense that the specific surface energy of the adhesive must be lower than that of the adherent. Accordingly, materials having high surface energy levels, such as metals, and those having medium surface energy levels, such as wood and paper, may be bonded relatively easily. In the case of polymers, bonding becomes increasingly more difficult [15], and finally almost impossible as surface energy decreases (polyolefins and polytetrafluoroethylene). The remedial surface treatment of materials such as these may be interpreted as increasing their surface energy [16].

The adsorption theory also shows that the adhesion forces of two materials are not reciprocal. For example, if a liquid epoxy resin adhesive is allowed to set on the surface of polytetrafluoroethylene or polyethylene, a very weak adhesive joint is formed. If, by contrast, liquid polytetrafluoroethylene or polyethylene is applied to the surface of hardened epoxy resin, strong adhesive joints are obtained [17]. In practice, this theory is not free from contradictions either; above all, it does not answer the question whether the difference in the surface energies between two materials is indicative of the intensity of the adhesion force [18].

Diffusion Theory. The diffusion theory of adhesion is essentially applicable to the bonding of high polymers. According to this principle, adhesion is obtained by the mutual penetration of adhesive and substrate [19], [20]. This mobility is based on the fundamental properties of high polymers: their chainlike structure and resulting mobility, allowing the possibility that the chains possess Brownian molecular movements in a submolecular range. By virtue of their greater mobility, the adhesive molecules normally play a greater part in the diffusion process. However, if the adhesive is in solution, which is generally the case, and if the substrate is slightly soluble in the solvent, substrate molecules or parts thereof also diffuse into the spread adhesive. A diffusion bond is characterized by the disappearance of a clear boundary between the two phases and by the development of a gradual transition from one phase to the other. The mechanism of adhesion has developed into a three-dimensional process and is no longer confined to one interface.

The interdiffusion of the polymer molecules of adhesive and substrate is dependent upon various parameters, such as pressure, time, temperature, molecule size, and, of course, the reciprocal solubility, as shown by the correlation between the compatibility of the polymers and the quality of the bond [21]. Examples of bonds to which the diffusion theory is applicable include the bonding of PVC-U adherents to PVC in solvents containing tetrahydrofuran and so-called contact bonding, where the diffusion process takes place between two adhesive layers and not between adhesive and substrate [22].

The limits of the diffusion theory of adhesion show up in the adhesion of polymers, for example, to metal or glass. In this case, this theory does not make any useful contribution to the understanding of adhesion.

Other Theories. In addition to these theories, some special cases are discussed in the literature. They include *adhesion by primary valence forces,* for example, in the bonding of metals [8], [23], and so-called *liquid adhesion.* In the latter, a thin film of a liquid of extremely high viscosity produces adhesion through a process in which separation of adherent and substrate results in a flow in the narrow gap which is only possible by overcoming considerable resistance. Liquid adhesion is particularly assumed in pressure-sensitive bonding [16] and in the initial tack of a liquid adhesive.

Summary. None of the theories covers every single aspect of bonding. A promising combination of the theories to bring about an improvement does not exist at present. In addition, the theory of adhesion as a physicochemical phenomenon and physically measurable adhesion are generally still a considerable distance apart, so that the adhesion theories are only reference points. Secondary factors, such as surface roughness, boundary layers with faults (atmosphere, impurities), joint design, type of stress, and aging also influence bond strength to a considerable extent.

1.3. Definitions

Some of the terms are defined in DIN 16920 [24] and EN 923 [132].

Adherent: a body that is or is intended to be bonded to another body.

Adhesion: state in which two surfaces are held together by interfacial bonds.

Adhesive application: the application and distribution of an adhesive over the surface to be joined.

Adhesive coat: adhesive layer applied to an adherent.

Appearance of separated surfaces: surfaces formed in the event of failure of an adhesive joint; they provide information on the reason for the failure. Distinctions can be drawn visually between:

1) A bond layer that does not cover the entire joint area (uneven adhesive application)
2) Adhesion failure: rupture of the bond layer from the surface of the adherent
3) Cohesion failure: break in the bond layer

4) Insufficient coalescence: separation of two adhesive layers at the interface, for example, in the case of contact adhesives
5) Rupture of the surface layer of an adherent, breakage of the adherent (material failure)

In many cases, several types of failure occur together.

Assembly time: interval between application of adhesive to the adherents and the initiation by heat and/or pressure of the setting process in the assembled joint.

Application weight, spread of adhesive, coverage: quantity of adhesive applied per unit area of a surface. A distinction is drawn between wet application (including solvent or dispersant) and dry application (100 % adhesive). Each type of adhesive has technically and economically an optimal application weight. In principle, rough and porous substrates (paper, fabrics) require more adhesive than smooth surfaces (glass, metals, films) to obtain whole-surface wetting and bonding. Application weights range from 1 g/m^2 (for example, for bonding films) to several 100 g/m^2, depending on the substrate, the method of application, and the type of adhesive.

Bond layer thickness: the thickness of the set layer of adhesive in millimeters. It may be influenced by the application weight and the fixing pressure. In general, a thin ($<$ 0.001 mm) and uniform bond layer leads to the highest bond strengths.

Bond line: adhesive layer in a bonded joint.

Bond strength: force necessary to bring an adhesive joint to the point of failure in or near the plane of the bond line.

Closed assembly time: interval between assembly of the adhesive joint and the initiation by heat and/or pressure of the setting process in the assembled joint.

Cohesion: state in which the particles of a single substance are held together by primary or secondary valence forces.

Cold flow: deformation of an adhesive layer (or film) at room temperature without external applied load.

Creep: slow inelastic deformation of an adhesive layer under load following initial instantaneous elastic or rapid deformation.

Curing: the setting of an adhesive by chemical reaction (polymerization, polycondensation, polyaddition).

Heat resistance: the maximum temperature that a set adhesive in a joint is capable of withstanding. It depends upon the nature and intensity of the load and must be determined in each individual case.

Initial tack: the ability of a setting adhesive to hold two substrates together immediately after they have been joined. At this stage of the bonding process, strength is determined largely by rheological properties and depends on the cohesion of the unset adhesive. In the case of hot-melt adhesives, the analogous term "hot tack" is used, and this property is determined by the viscosity of the unsolidified melt. The transitions between the use of the terms initial tack and tack are fluent.

Minimum film-forming temperature: the temperature at which an emulsion-based adhesive just no longer dries to form a clear, homogeneous film.

Open assembly time, open time: interval between adhesive application to the adherents and assembly of the adhesive joint. The expression "open time" also is used occasionally for the period of time that elapses in a machine between application of the adhesive and bonding.

Primers: (mixtures of) substances that are applied before the adhesive to one or both of the surfaces to be joined, for improving adhesion and/or durability of the bond.

Setting: process by which an adhesive develops its cohesive strength and hence the physical and chemical properties of its bond.

Setting rate: the increase in strength in the bond line per unit time. It depends upon the setting mechanism and temperature and also may be influenced by the substrate (absorbency), bond layer thickness, etc. The setting rate generally is not constant and decreases toward the end of the setting time.

Setting time: time between joining of the adherents and the setting of the adhesive. It varies from a few seconds to several days, depending on the type of adhesive.

Shear strength: force per unit surface area necessary to bring an adhesive joint to the point of failure by means of shear forces.

Shelf life, storage life: time of storage under stated conditions during which an adhesive can be expected to retain its working properties. Appropriate storage generally is understood to mean that the adhesives are stored in clean, hermetically sealed containers and in cool, dry, well-ventilated rooms. Shelf life is largely dependent upon temperature. Aqueous and in some cases even solvent-containing systems have a tendency for skin formation, separation, and an increase in viscosity, particularly when stored for long periods at relatively high temperatures. They also can be spoiled by microbial attack. Wide fluctuations in temperature (freeze/thaw) lead to coagulation. Shelf lives of less than 12 months generally are indicated on the packing. Manufacturer's instructions should always be followed.

Tack: property of a material that enables it to form a bond immediately on contact with another surface, which can be an adherent or another layer of adhesive. High tack is particularly important in the case of pressure-sensitive adhesives (tapes, labels), for which it is even measured. Tack is generally difficult to measure in setting adhesives.

Thixotropy: decrease in apparent viscosity under shear stress, followed by a gradual recovery when the stress is removed. Thixotropic behavior allows pasty adhesives, for example, to be spread readily and applied in relatively thick layers, in some cases even on vertical surfaces, without running or flowing out of the gap before bonding because the adhesive loses its fluidity immediately after application.

Viscosity: property of a material to increasingly resist deformation with increasing rate of deformation. This property is quantitatively defined as dynamic viscosity or coefficient of viscosity and is often used synonymously with apparent viscosity. The viscosity of adhesives is primarily determined by means of rotational or throughflow viscometers (DIN cup, Ford cup, Zahn cup). Adhesives generally show non-Newtonian behavior. In addition to temperature, any expression of viscosity must also refer to the measuring instrument and measurement parameters (rotating spindle, rate of shear, nozzle diameter).

Working life, pot life: period of time during which a multicomponent adhesive can be used after mixing the components. The pot life of a given system depends primarily on the initial temperature of the mixture and on the quantity used (a high temperature and large quantities result in shorter application times), although it also depends on the mixing conditions (with or without cooling, type of stirrer used), on the ratio of surface area to volume of the mixing vessel, and on the method used for end-point determination. Moisture-curing single-component adhesives also can have a working life, which begins with the contact of atmospheric moisture with the adhesive and which, apart from temperature, is largely dependent upon the existing humidity and upon the degree of moisture uptake during stirring operations.

Standards: comparisons of ASTM, MIL, GB, ISO, and DIN standards can be found in [25] and [26].

Legal Regulations. In view of the formulation of adhesives, particularly the solvent content, their manufacture, packaging, storage, transportation, and application are governed by a large number of regulations, which concern generally goods containing inflammable, explosive, or physiologically harmful constituents. Clean-water and clean-air regulations also must be observed.

1.4. Raw Materials for Adhesives

The adhesives industry uses many different products of natural and synthetic origin. These are mainly raw materials that are produced in large quantities and hence at reasonable prices. Raw materials that are used solely or primarily for adhesives are relatively rare or, alternatively, represent special types within fairly broad production programs.

Basic raw materials are natural and synthetic polymers or monomers and prepolymers which can form such polymers. The polymers primarily impart the required strength (cohesion) to the adhesive layer. In many cases, however, they have adequate adhesion properties of their own. The adhesion properties may be improved by addition of resins or special adhesion promoters. Plasticizers and resins improve the flexibility of the polymers, increase the tack of the adhesives, or establish other required product properties, e.g., rheological properties. Rheological and other special properties can be influenced by adding fillers. Polymers are often dissolved in water or an organic solvent to achieve the wetting necessary for good adhesion.

The required service properties frequently can be obtained by using different basic raw materials or different kinds of formulations. Accordingly, the raw materials often cannot be assigned unequivocally to specific types of adhesive. Because almost every polymer and resin and also many other substances in principle may be used for the production of adhesives, it is not possible to provide a complete list of every possible raw material. For this reason, only the most important are listed in the following.

Starch [*9005-25-8*]. Although their significance has decreased, the various types of starch, e.g., mainly potato and wheat starch in Europe, corn starch in North America, and tapioca in Latin America, Africa, and Asia/Pacific region, still represent an important group of raw materials in quantitative terms. Starches are used in native form, in degraded, hot- or cold-soluble form, and as solutions.

In addition, they are used in combination with casein solutions and synthetic resin emulsions. Whereas starch ethers and esters and also aldehyde starches are used to only a limited extent in the adhesives field, the thermal degradation products of starches, the water-soluble dextrins, are still important in quantitative terms as raw materials for adhesives. They are used on their own in aqueous form or as mixtures with synthetic resin emulsions.

Proteins. Whereas vegetable protein is of virtually no significance, milk protein is still important, although increasingly less so, for casein glues. The same applies to collagen from animal skins and bones for glue jellies and glutins.

Cellulose Ethers and Cellulose Esters. Cellulose ethers, primarily methyl cellulose, are used as aqueous solutions (wallpaper paste) and also as thickeners.

Nitrocellulose [*9004-70-0*] is used as a raw material for solvent adhesives (multipurpose adhesives, shoe adhesives), although in recent years it has been increasingly overtaken in importance by other polymers.

Natural Rubber [*9006-04-6*]. Pressure-sensitive adhesives, contact adhesives, rubber solutions, and other solvent adhesives are made from natural rubber (generally first-latex crepe). Solutions of regenerated rubber also have some significance. Before dissolution, natural rubber generally has to be degraded by mastication on rolls or in kneaders and solubilized. Recently, soluble types also have been available. Natural rubber latices (rubber latex) having solids content of from 35 to 70 % are widely used for special latex adhesives (envelope gumming and the like). Compared with synthetic rubber, natural rubber has better tack but is more sensitive to heat and oxidation, particularly in contact with nonferrous metals.

Synthetic Rubber. Many different types of synthetic rubber are suitable raw materials for adhesives and sealing compounds. Particularly significant are polychloroprene rubber, styrene – butadiene rubber (SBR), nitrile rubber, and polyisobutylene. Unless these rubbers are available as directly soluble types, they have to be degraded by mastication on rolls or in kneaders and solubilized before dissolution.

Standard contact adhesives and pressure-sensitive adhesives are made from SBR in solution. Styrene – butadiene rubber latices, in some cases containing carboxyl groups, are used for special emulsion-based adhesives. Thermoplastic elastomeric block copolymers of styrene with butadiene or isoprene are of increasing significance in contact adhesives and hot-melt adhesives.

Polychloroprene rubbers are the principal raw material for high-quality contact adhesives. Grades with high degrees of crystallinity are used preferably. Polychloroprene latexes are raw materials for aqueous contact adhesives.

Nitrile rubber is used preferably for contact adhesives with improved plasticizer resistance. Polyisobutylene is used in pressure-sensitive adhesives. Butyl rubber is sometimes added to pressure-sensitive and hot-melt adhesives, although it is mainly used in sealing compounds. Epoxy resins and reactive (meth)acrylate adhesives (reactive adhesives) are modified with polychloroprene, butyl, and nitrile rubber. Chlorinated rubber is added in small quantities to contact adhesives and also to rubber-to-metal bonding agents for improving the adhesion properties.

Polyethylene [*9002-88-4*]. Low molecular mass polyethylene, vinyl-modified polyethylene and polyethylene waxes are used for hot-melt adhesives.

Chlorosulfonated polyethylene can used in reactive adhesives in combination with (meth)acrylate monomers and also in rubber-to-metal bonding agents.

Ethylene – vinyl acetate copolymers are the most important polymer raw materials for hot-melt adhesives. They contain 18 – 45 % vinyl acetate and have melt flow indices of from less than 1 to 2500. The quantitatively most important copolymers contain ca. 30 % vinyl acetate and having melt flow indices of 40 – 400. Ethylene – vinyl acetate copolymer emulsions containing ca. 20 % ethylene are used in emulsion-based adhesives.

Polypropylene [*9003-07-0*] and various copolymers are used in atactic form for standard hot-melt adhesives.

Poly(vinyl esters) are one of the most important groups of raw materials for adhesives. Poly(vinyl acetates) are important in solvent adhesives. Synthetic resin emulsions, based on poly(vinyl esters) with a solids content of 50 – 70 % are the principal raw materials for emulsion-based adhesives.

Besides poly(vinyl acetate) emulsions, emulsions of vinyl acetate copolymers with ethylene, (meth)acrylates, vinyl chloride, maleic esters and vinyl laurate are of considerable importance as internally plasticized synthetic resin emulsions.

Poly(vinyl acetate) emulsions for adhesives contain preferably poly(vinyl alcohol) as a protective colloid. Spray-dried poly(vinyl acetate) emulsions, known as redispersion powders, are used in combination with cellulose ethers as binders in floor-leveling compositions and for increasing solids content in emulsion-based adhesives. Poly(vinyl acetates) are used as solid polymers for formulating solvent adhesives.

Poly(vinyl chloride) [*9002-86-2*] homopolymers are processed primarily into plastisol adhesives and special solvent adhesives. Chlorinated PVCs (ca. 65 % Cl) are more readily soluble and are used for certain solvent adhesives.

Copolymers of vinyl chloride are used both as solid resins, for example, in solvent adhesives (vinyl chloride – vinyl acetate copolymers), and as additives in plastisol adhesives, but especially as synthetic resin emulsions for heat-sealing adhesives. Important copolymers are those of vinyl chloride with vinyl acetate, acrylates, or maleates, as well as copolymers of vinylidene dichloride with acrylates or acrylonitrile.

Poly(vinyl alcohol) [*9002-89-5*] is used primarily as a protective colloid for poly(vinyl acetate) emulsions. Relatively small quantities are added to modify emulsion-based adhesives. Aqueous solutions of poly(vinyl alcohol), in some cases combined with fillers and emulsions, are used as paper glues and as labeling adhesives.

Poly(vinyl acetals). Poly(vinyl formal) and poly(vinyl butyral) are used in solution as primers, in combination with phenolic resins for specialty reactive adhesives, and as films for the manufacture of laminated glass.

Acrylate and Methacrylate Polymers. Poly(ethyl acrylate) and poly(butyl acrylate) solutions and emulsions are important raw materials for pressure-sensitive adhesives. Copolymers of various esters, which give films of tailor-made hardness and which may additionally contain functional groups (carboxyl, amide, amino, methylol, hydroxyl), are used for pressure-sensitive adhesives to improve the adhesion properties or to enable the adhesive layer to be cross-linked to a limited extent.

Methyl acrylates are used as comonomers for increasing the hardness of the adhesive layer. Acrylate – styrene copolymers are used as emulsions for coating paper.

Solvent adhesives and reactive adhesives are made from homo- and copolymers of methacrylates, generally methyl and ethyl methacrylate and, occasionally, butyl methacrylate. Monomeric (meth)acrylates are also used in reactive adhesive systems (polymerization adhesives). Poly(ethylene glycol) dimethacrylates are the basis of anaerobically curing liquid resins (reactive adhesives). They also are added as adhesion promoters to plastisol adhesives. Acrylate – ethylene copolymers, in some cases with a small content of carboxyl groups, are used instead of ethylene – vinyl acetate copolymers as fusible polymers for special hot-melt adhesives. Salts of polyacrylate and acrylate – acrylic acid copolymers are used as thickeners for aqueous adhesive solutions and emulsion-based adhesives.

α-Cyanoacrylic esters (above all methyl, ethyl, butyl, but also allyl and methoxyethyl esters) are used for one-component reactive adhesives.

Poly(vinyl ethers) are used for specialty pressure-sensitive adhesives.

Poly (vinyl pyrrolidone) [*9003-39-8*] is used in gluesticks and in aqueous pressure-sensitive adhesives.

Polystyrene [*9003-53-6*] is used as a solution for bonding polystyrene and other plastics.

Polyamides and Polyamidoamines obtained by the condensation of dimerized fatty acids with aliphatic diamines are used in solid form as hot-melt adhesives with particularly pronounced adhesion properties. The low molecular mass condensates, some of which are still liquid and contain free amino groups (amine equivalent ca. 3000 – 12 000), are used widely as hardeners for epoxy resins.

Copolyamides based on C_{12} and C_{14} amino acids or lactams are preferably used as heat-sealing powders for textile applications.

Polyesters. Unsaturated polyesters dissolved in styrene or methacrylates are used as reactive adhesives. Hot-melt adhesives can be made from saturated and fusible polyesters based on aliphatic and/or aromatic carboxylic acids, in some cases by using dimerized fatty acids. High molecular mass amorphous polyesters and low molecular mass crystalline polyesters are important raw materials for moisture cross-linking polyurethane adhesives in solution or solvent-free systems.

Low molecular mass polyesters and polycaprolactones containing hydroxyl groups are important raw materials for polyurethane adhesives.

Polyurethanes, by virtue of their considerable variability, have become very valuable to the adhesives industry over recent decades. They generally show above-average adhesion and cohesion properties. Low molecular mass reaction products of hydroxyl compounds and polyisocyanates, which contain cross-linkable hydroxyl or isocyanate groups, are used for various reactive adhesives. The hydroxyl components are aliphatic and aromatic polyesters, polyethers, castor oil, and even simple polyalcohols.

High-polymer polyurethane rubbers with various crystallization properties are used for solvent adhesives, in some cases in combination with resins and other polymers, or for contact adhesives. In many cases, solutions of this type are also used for heat-reactivatable adhesive coats. By adding polyisocyanates, it is possible to improve both the adhesion properties and the bond strength at elevated temperature.

Polyisocyanates. Whereas any standard, generally difunctional, polyisocyanate, but preferably toluene diisocyanate (TDI) and methylenediphenyl diisocyanate (MDI), may be used for the production of polyurethane resins, only polyisocyanates with low vapor pressures at room temperature, such as MDI and its homologues, the reaction product of trimethylol propane with TDI, and triphenylmethane-4,4′,4″-triisocyanate, are recommended as hardeners for adhesives for reasons of industrial hygiene.

Nonaromatic isocyanates such as isophorone diisocyanate (IPDI), hexamethylene diisocyanate prepolymers, and dicyclohexylmethane diisocyanate prepolymers have also gained some importance. Guidelines for the use of isocyanates are given in [27].

Epoxy resins are used primarily in combination with hardeners (polyamines, polyaminoamides, dicyanodiamide, acid anhydrides) for cold-setting and hot-setting reactive adhesives. Liquid epoxy resins of the bisphenol A type with epoxy equivalents of 170 – 300 are used preferably; relatively high molecular mass solid resins are used only occasionally.

In quantitative terms, novolak epoxy resins (epoxy equivalent 120 – 200) and elasticized or otherwise modified bisphenol A (F) resins have relatively little significance.

Epoxy resins are also combined with phenolic resins and poly(vinyl acetate) resins and used as adhesion promoters, for example, in plastisol adhesives.

Phenolic Resins [*9003-35-4*]. Curable phenol – formaldehyde resins of the resol type are of considerable importance as hot-setting reactive adhesives for bonding wood and metals. Suitable types are used also in combination with poly(vinyl acetals) and epoxy resins for special, in some cases filmlike, reactive adhesives for bonding metals.

Standard and modified phenolic resins, alkyl phenol resins, etherified phenolic resins, and phenol – terpene resins also are used as primers and resin components in other types of adhesives, for example, in contact cements, pressure-sensitive adhesives, rubber-to-metal bonding agents, and similar adhesive systems.

Resorcinol Resins [*24969-11-7*]. Resorcinol – formaldehyde resins and resorcinol – phenol – formaldehyde resins are used as cold-setting reactive adhesives for waterproof wood bonds and as binders between tire cord and rubber.

Urea Resins [*9011-05-6*]. In quantitative terms, urea–formaldehyde resins are one of the most important groups of raw materials for adhesives. They are used in the form of aqueous solutions or spray-dried powders, generally containing acid hardeners, as reactive adhesives for bonding wood, particularly for the production of plywood and chipboard.

Modified urea resins are also important for other purposes, for example for labeling adhesives and as cross-linking components.

Melamine–formaldehyde resins [*9003-08-1*], like urea resins, are used primarily for bonding wood. Because they are not particularly stable in aqueous solution, melamine–formaldehyde resins are marketed primarily in powder form.

Nonreactive resins with a variety of compositions are of considerable importance in the formulation of adhesives. In some cases, they are used on their own in solvent adhesives, so-called resin adhesives, but generally in combinations with polymers, in which they perform various functions. Above all, they increase tack, improve adhesion, influence viscosity, fluidity, and sealability, and, in some cases, also act as plasticizers. The most important applications for nonreactive resins are in pressure-sensitive adhesives, contact adhesives, hot-melt adhesives, solvent adhesives, and emulsion-based adhesives. In this context, the term resin covers materials differing very widely in their composition:

- Rosin and its derivatives
- Tall-oil rosins and their derivates
- Hydrocarbon resins
- Carbamic ester resins
- Cyclohexanone resins

Plasticizers are used in adhesive formulations, above all for elastification of the polymer films in emulsion-based adhesives and also for increasing wet tack. They are used, e.g., in homopolymer emulsion-based adhesives. Phthalate plasticizers are the most widely used, diisobutyl phthalate being especially important for emulsion-based adhesives. In addition to phthalate and adipate plasticizers epoxy and phenol sulfonic amide plasticizers are used in PVC plastisols. Besides ester plasticizers, mineral oils and also chlorinated low molecular mass polyethylenes and low molecular mass hydrocarbon resins are used in rubber adhesives.

Other plasticizers, such as benzoates, citric esters, and glycerol triacetate are less important. In hot-melt adhesives, mineral oil and low molecular mass polyolefins are used.

Solvents, primarily for solvent adhesives, but sometimes also added to emulsions in small quantities as temporary plasticizers, are mainly petroleum, toluene, xylene, dichloromethane, trichloroethylene, trichloroethane, ethyl acetate, acetone, 2-butanone, methyl isobutyl ketone, and cyclohexanone. The use of solvents has been reduced significantly by the introduction of solvent-free adhesives (e.g., polyurethanes) in several market segments. Aromatic solvents, such as toluene and xylene, and chlorinated products like trichlorethane and dichloromethane have been nearly eliminated or are used in closed systems.

Fillers are added to adhesives primarily for establishing certain consistencies and occasionally for reducing cost. Preferred fillers are pyrogenic and precipitated silicas, chalks, and light and heavy spar. Fibers and metal powders are used in special cases [28].

Other Additives. Small quantities of standard preservatives, antioxidants, UV/light stabilizers, colored pigments, and defoaming agents are added to many adhesives.

1.5. Classification of Adhesives

Adhesives can be classified on the basis of chemical composition [29], setting mechanism [30], and adhesion mechanism as pressure-sensitive, adhesion, and diffusion adhesives [31]. LUCKE [32] proposed a flexible classification in which each adhesive is characterized and classified according to various characteristic features, such as chemical basis, form of application, application temperature, thermal behavior, and uses. A classification such as this provides for very detailed coverage of the individual features of an adhesive and is a helpful guide for industrial users. However, it is extremely comprehensive and goes beyond the scope of this treatise. For this reason, a classification based on the setting mechanism is given here [30].

1. Setting without Chemical Reaction. The high molecular mass bonding substance is present prior to bonding.

1.1. Application without Volatile Solvents. The adhesive film is obtained by the action of heat.

Hot-melt adhesives, e.g., ethylene–vinyl acetate copolymers, polypropylene, polyamides, polyesters. Uses: paper, fiberboard, plastics, textiles, leather (assembly bonds).

Plastisol Adhesives. PVC or poly(methyl methacrylate) powder + plasticizer + adhesion promoter. Uses: metals (sheet-metal constructions), silicate-containing materials.

1.2. Application of Solutions, Solvent Escapes before Bonding. Adhesive layer is heated before or during bonding.

Heat-Sealing Adhesives and *High-Frequency Welding Auxiliaries,* e.g., vinyl polymers. Uses: paper, plastics, lamination of plastic films to metal foils (packaging, metal-foil lamination).
Contact Cements: elastomers (e.g., polychlorobutadiene) in conjunction with adhesive resins (e.g., alkylphenol or modified phenolic resins); bonding is obtained by contact under pressure between two almost dry adhesive layers. Uses: wood, plastics, rubber, metals.
Pressure-Sensitive Adhesives, e.g., natural and synthetic rubbers in conjunction with modified natural resins and/or synthetic resins; also polyacrylates, poly(vinyl ethers). Uses: nonspecific adhesion to virtually any surfaces, more particularly for the permanently tacky coating of tapes, films, and labels.

1.3. Application of Solutions, Solvent Evaporates during Bonding.

Solutions of polymeric, natural, and/or synthetic substances in organic solvents, e.g., vinyl polymers, natural rubber, synthetic rubbers of low polarity. Uses: paper, plastics, silicate-containing materials, wood.
Solutions of natural and synthetic high molecular mass substances in water, e.g., starch, dextrins, casein, cellulose ethers, water-soluble derivatives of poly(acrylic acid), poly(vinyl alcohol), poly(vinyl pyrrolidone) (adhesive sticks). Uses: paper, fiberboard.
Glutins (glues of animal origin). Uses: wood, paper, fiberboard, moistenable adhesive tapes.

1.4. Emulsions of Water-Insoluble Substances in Water that Escapes during Bonding.

Aqueous emulsions of polymeric compounds, e.g., homopolymers of vinyl acetate and vinyl propionate, vinyl acetate copolymers with ethylene or maleic esters; polyacrylic esters, styrene copolymers. Uses: paper, wood, also bonding to plastics.

2. Setting by Chemical Reaction (Reactive Adhesives). The high molecular mass bonding substance is formed during bonding.

2.1. Formation and Cross-linking of the Adhesive Film by Polymerization.

Two-pack polymerization adhesives: unsaturated polyesters with monomeric vinyl compounds, e.g., styrene or methyl methacrylate. No-mix adhesives and A/B acrylates, applied without premixing. Uses: metals, plastics, silicate-containing materials.
One-pack polymerization adhesives:

a) Cyanoacrylate adhesives setting under the influence of moisture or alkalinity. Uses: metals, plastics, silicate-containing materials, rubber, wood (high-strength spot bonds or small-area bonds).
b) Anaerobic adhesives. Special methacrylate adhesives. Uses: metals (threadlocking and bearing retention).

2.2. Formation and Cross-linking of the Adhesive film by Polyaddition.

Epoxy resin adhesives, polyepoxy compounds in combination with amines, polyamidoamines, or dicarboxylic acid anhydrides. Uses: metals, silicate-containing materials, plastics.
Reactive polyurethane adhesives: polyurethane prepolymers with terminal hydroxyl groups or terminal isocyanate groups. Uses: plastics, metals, silicate-containing materials.
Two-component silicone adhesives: setting induced by catalyst.

2.3. Formation and Cross-linking of the Bond by Polycondensation (elimination of low molecular mass cleavage products during setting).

Polyhydroxymethyl compounds: phenol – formaldehyde resins, also in combination with poly(vinyl formal) resins, nitrile rubber or epoxy resins (as adhesive film), urea – formaldehyde resins, melamine – formaldehyde resins, resorcinol – formaldehyde resins. Uses: wood (plywood manufacture), metals.
Silicone adhesives, one-pack, setting under the influence of atmospheric moisture. Uses: glass, metals, ceramics, plastics; some substrates need a priming coat.
MS polymers, silane-alkoxy-modified prepolymers of poly(propylene glycol), one-pack systems. Uses: Many surfaces; no primer necessary.
Polyurethane adhesives, one-pack, setting under the influence of atmospheric moisture. Uses: plastics, metals, silicate-containing materials.
Reactive polyurethane hot-melt adhesives, one-pack, setting under the influence of atmospheric moisture. Uses: bookbinding (high-quality perfect bindings), wood gluing, shoe manufacturing.
Polyimides and polybenzimidazoles. Uses: metals (for high thermal stressing).

2.4. Cross-linking of Polymeric Adhesive Components in the Bond. Rubber-to-metal bonding agents, e.g., halogenated polymers in conjunction with cross-linking agents; adhesion is obtained under vulcanization conditions (elevated temperature and pressure). Uses: natural and synthetic rubber to metals (structural elements for damping vibration).

2.5. Ultraviolet/Electron Beam (UV/EB) Curing Adhesives. Special acrylic acid esters (monomers or oligomers). Uses: for laminating and for pressure-sensitive products.

2.6. Conductive adhesives include electrically and thermally conductive materials. Most fillers that impart electrical conductivity also contribute to thermal conductivity, but thermally conductive adhesives that are electrically insulative are also available.

1.6. Individual Adhesive Systems

The same classification used in Chapter 1.5 is used here. Information on the uses of the adhesives discussed here can be found in Chapter 1.5.

1.6.1. Adhesives That Set Without a Chemical Reaction

1.6.1.1. Solvent-Free Adhesive Systems

Hot-melt adhesives are 100 % solids that, in the broadest sense, include all thermoplastic polymers. Polymers that are primarily used as hot-melt adhesives include ethylene – vinyl acetate copolymers (EVA), polyvinyl acetates (PVA), polyethylene (PE), amorphous polypropylene, block copolymers (thermoplastic elastomers), polyamides, and polyesters. The oldest hot-melt adhesive, which has been in use since early times, is sealing wax. In principle, glutins and glue jellies also may be regarded as hot-melt adhesives. However, modern hot-melt adhesives are primarily synthetic products. The simplest hot-melt adhesives are rosin – wax mixtures. But these products have limited strength and thermal stability.

In general, hot-melt adhesives are solid below 80 °C. Ideally, as the temperature is increased beyond this point, the material rapidly melts to a low-viscosity fluid that can be easily applied. Upon cooling, the adhesive sets rapidly. Because these adhesives are thermoplastics, the melting/resolidification process is repeatable. Typical application temperatures of hot-melt adhesives are 150 – 190 °C with melt viscosities in the range of 500 – 3000 mPa s.

Hot-melt adhesives do not undergo a chemical reaction because they simply cool from the melted state to form a solid film at the bond line. These adhesives are available in forms such as pellets, slabs, bars, slugs, and films that allow convenient handling by a variety of application equipment. Because hot-melt adhesives are applied in the molten state, melt viscosity is an important property. The limitations of the application equipment often influence the viscosity range that is selected for a hot-melt adhesive formulation.

Although hot-melt adhesives consist of 100 % solids, they are rarely 100 % polymers in composition. This is due to the limited adhesion of pure thermoplastics like EVA copolymers and their lack of melt properties such as tack and wettability. The components of a hot-melt adhesive can be roughly divided into two categories: polymers and diluents. Typically, diluents are waxes, plasticizers, tackifiers, stabilizers, extenders, and pigments. The functions of the diluents are:

- Lowering the viscosity for easy application
- Enhancing wettability
- Enhancing adhesive strength

– Increasing rigidity (extenders) or flexibility (plasticizers)

It is desirable to have a hot-melt adhesive with high strength. But this property is generally accompanied by a high melt viscosity. A low melt viscosity is desirable for application purposes. Therefore hot-melt adhesive formulations must include diluents that decrease the polymer melt viscosity and increase the overall adhesive strength.

Special dispensing equipment must be used to apply hot-melt adhesives. Typical application methods include the use of rollers, screw extruders, and squirting pumps or nozzle applicators.

The polymers are generally of high molecular mass and impart strength and high viscosity to an adhesive. Waxes such as paraffin wax can function as both diluents and antiblocking agents. In addition, waxes promote surface wetting. Similarly, plasticizers such as phthalates, mineral oils, and glycolates provide both surface wetting and adhesive flexibility. Typical tackifiers include rosin, modified rosin, terpenes, modified terpenes, hydrocarbons, and chlorinated hydrocarbons. These materials provide tack and flexibility while promoting surface wetting and adhesion. Stabilizers such as hindered phenols help to maintain adhesive melt viscosity while functioning as antioxidants. Extenders such as talc, clay, and barites lower the cost of hot-melt adhesives and simultaneously help control melt flow.

The following properties of hot-melt adhesives are important:

Melt viscosity is one of the most important properties of a hot-melt adhesive. In general, as the temperature of a polymer increases, its viscosity decreases. Therefore, in a hot-melt adhesive formulation, the melt temperature controls the viscosity, which greatly influences the extent of surface wetting. The temperature of the melt and the application equipment should be maintained as constant as possible.

The *bond formation temperature* is the minimum temperature below which surface wetting is inadequate. A hot-melt adhesive is applied at a running temperature at which the viscosity is sufficient to wet surfaces properly. Running temperatures that are too high allow more time for surfaces to be wet properly. However, if low penetration is required and if the substrates are porous, this longer time may be detrimental and may result in a bond that contains insufficient adhesive. A running temperature that is too close to the bond-formation temperature may result in a too-rapid solidification of the adhesive bond. As a result, overall properties such as strength may not be optimized.

The *heat stability* of the adhesive is a very important property. Running temperatures that are greater than the adhesive degradation temperature result in charring and decreased overall properties. The addition of stabilizers to the adhesive formulation contributes to the stability of the material by hindering the acceleration of degradation due to the presence of oxygen.

The *tack* of the adhesive indicates the stickiness of the hot-melt as it changes from a liquid to a solid state. This property affects the ability of the adhesive to hold the substrates together.

The *open time* is the lapse of time between the application of the hot-melt to one substrate and the loss of its wetting ability on the second substrate due to solidification.

It typically ranges from fractions of seconds to minutes to infinity (in pressure-sensitive hot-melt adhesives).

The *limitations* of hot-melt adhesives are restricted toughness at usable viscosities, low heat resistance, and poor creep resistance. All properties are affected by the polymer and diluents used in the formulation. The application conditions, including the amount of adhesive and the pressure applied to the bond line, also affect end-use properties. In the hot-melt adhesive-bonding operation, a minimum amount of pressure must be applied until the hot-melt becomes solid or sufficient tack develops to hold the substrates in place.

Manufacture. In the manufacturing of hot-melt adhesives, the components are melted in heated stirring vessels and subsequently formed in pelletizing or granulating machines. The melt is allowed to drop onto a cooling belt or to form a film on a cooling belt and subsequently cut. Another possibility is extrusion from an extruder followed by underwater granulation. For manufacturing hot-melt adhesives with higher viscosity and higher filler contents, it is preferable to use heatable kneaders or extruders followed by specialized processing units. Hot-melt sticks (slugs, rods) for application by handguns are produced by injection molding or by cutting hot-melt profiles.

Surface-tacky hot-melt adhesives are marketed in containers or in the form of blocks or cubes in readily removable packs (silicone-coated paper) [33] or in meltable polyethylene bags.

Hot-Melt Chemical Families. EVA copolymers are the cheapest hot-melt adhesives and are used in the greatest quantities. Typical properties of these adhesives include sufficient strength between ca. 30 and 50 °C, but limited upper service temperature of 60 – 80 °C, low hot-melt viscosity, and low creep resistance under load with time. Most EVA-based hot-melt adhesives are used to bond paper, fabrics, wood, and some thermoplastics.

PP polymers based on amorphous polypropylene have better properties than EVA hot-melt adhesives. In particular, higher heat resistance and better adhesion properties can be obtained.

Polyamides and thermoplastic polyesters are classified as high-performance hot-melt adhesives. These adhesives have greater strengths than EVA-based hot-melt adhesives and exhibit adequate bonding from approximately – 40 to 70 °C. Some formulations have upper service temperatures up to 185 °C for applications that do not involve loads. One disadvantage of the higher service temperatures is the higher hot-melt application temperature, which may require special equipment. Both polyamides and polyesters are sensitive to moisture absorption during application. This can result in hot-melt foaming, which leads to voids in the solidified adhesive and decreased strength. Nonetheless, polyamides exhibit good strength and flexibility. Polyesters are even stronger, but are more rigid and are usually highly crystalline. This results in a sharp melting temperature, which is desirable for high-speed bonding. Polyamide and polyester hot-melt adhesives are used to bond plastics, glass, wood, leather, foam, fabric, rubbers and some metals, such as aluminum, copper, and brass.

Polybutene-based hot-melt adhesives are tough, partially crystalline, and their slow crystallization rates lead to long open times. Copolymers of butene result in softer and more flexible adhesives. In general, polybutene and its copolymers have low temperatures for recrystallization from the melt. This permits stress release in the adhesive bond, which may have been applied to cold surfaces. Polybutene and its olefinic copolymers exhibit good bonding to nonpolar surfaces but poor compatibility with polar substances. These hot-melt adhesives have been used on rubbery substrates and are available as pressure-sensitive adhesives.

Thermoplastic elastomers that are formulated into hot-melt adhesives include polyurethane and block terpolymers, such as styrene – butadiene – styrene; styrene – isoprene – styrene; and styrene – olefin – styrene, in which the olefin component is typically ethylene, propylene, and/or butylene. The saturation in the mid-segment of these terpolymers results in better UV and thermooxidative resistance than that of unsaturated butadiene and isoprene midsegments.

In general, thermoplastic elastomer hot-melt adhesives are not as strong as polyesters. However, they exhibit good flexibility, high elongation, toughness, and vibration resistance. Polyurethanes and block terpolymers are used as hot-melt pressure-sensitive adhesives in tapes and labels and for nonwoven applications.

Pressure-sensitive hot-melt adhesives are surface-tacky and can be used by the application of pressure alone. These adhesives have infinite open times and are used at room temperature. Typically, pressure-sensitive hot-melt adhesives are supplied in form of blocks or cubes in silicone coated packaging units or preferably in polyethylene bags, which after melting form part of the adhesive.

Foamable hot-melt adhesives have been commercially available since 1981. Nitrogen or carbon dioxide is introduced into the hot-melt adhesive and results in a 20 – 70 % increase in adhesive volume. Foaming increases hot-melt spreading and open time. This method is usually used with polyethylene hot-melt adhesives that are applied to selected metal, plastic, paper, porous, and heat-sensitive substrates.

Low-melt adhesives are hot-melt adhesives with an application temperature in the range of 100 – 120 °C instead of 160 – 190 °C. They are used for bonding heat-sensitive materials.

Plastisol adhesives are solvent-free adhesives that require temperatures of 120 – 200 °C for setting. They consist essentially of a dispersion of finely divided poly(vinyl chloride) or polymethacrylate [34] in plasticizers and low molecular mass, heat-reactive substances that act as adhesion promoters, for example, epoxy resins, poly(ethylene glycol) dimethacrylates, phenolic resins; in some cases these are even used in conjunction with corresponding hardeners. Standard heat stabilizers for poly(vinyl chloride) are also added to prevent decomposition of the poly(vinyl chloride) and evolution of HCl during hardening or in the event of subsequent exposure to heat. Depending on the end use, varying amounts of fillers, pigments, or thixotropizing agents may be added to modify the color and fixing behavior. At a certain setting or hardening temperature, the poly(vinyl chloride)/polymethacrylate plasticizer mixture becomes a dispersion gel and also adheres to the substrate through reaction of the adhesion promoter. The various adhesion promoter systems require different activation temperatures, usually in the range of 120 – 200 °C [34].

Production and Storage. Because of their relatively high viscosities, these pasty to highly viscous compositions must be produced in kneader-type stirring vessels capable of applying considerable forces. It is important to ensure that stirring and kneading do not generate excessive temperatures, which can result in gelation of the product. For the same reason, the products must be stored under cool conditions (below 30 °C).

1.6.1.2. Adhesive Solutions from Which the Solvents Escape before Bonding

This group includes heat-sealing adhesives and high-frequency (HF) sensitive heat-seal coats; these are coating adhesives that are heat-activated during bonding. Contact adhesives and pressure-sensitive adhesives bond by the same principle but without the action of heat.

Heat-sealing adhesives are applied to the materials to be sealed in the form of solutions, emulsions, or, preferably, in melt or powder form. The solvent-free layer is intended to be nontacky and only to melt and to wet the adherent under the action of heat during sealing. Solidification takes place after cooling. In many cases, sealing is carried out on a coating-to-coating basis.

The choice of raw materials for sealing coatings is very wide and is influenced by the sealing temperature and the sealing behavior of the materials used, by the requirements the joint has to satisfy, and by the adhesion of the sealing layers to the substrates. In many cases, copolymers of vinyl chloride or vinylidene chloride are used in solution, although they are also used in emulsion in conjunction with resins and other polymers. In addition, copolymers of vinyl acetate and polymethacrylates, polyurethanes, and polyesters are also used. Ethylene – vinyl acetate copolymers are used preferably for heat-sealing coatings applied as melts. Heat-sealing powders also are used occasionally; they can be based on fine polyamide or polyethylene powders and also on vinyl acetate – vinyl chloride copolymers.

High-frequency-sensitive heat seal coats are a special type of heat-sealing adhesive [35]. They are also applied in the form of a solution or emulsion to the surface of a substrate and dried. However, welding to the partner substrate is not obtained by the application of heat through the substrate; instead the sealing temperature is generated by high frequency within the adhesive layer itself. Hence, the adhesive layer must have a high dielectric loss factor.

High-frequency sealing auxiliaries generally contain vinyl chloride copolymers in addition to proportions of other polymers, such as polyacrylates, vinyl acetate copolymers, plasticizers, and resins. High-frequency heatable coatings have solids contents of from 30 % (solutions) to 50 % (emulsions). Their activation temperature is in the range of 120 – 150 °C. High-frequency sealing powders consist predominantly of finely divided polyamides.

Contact cements must be applied to both substrate surfaces to be joined. The adhesive layers are joined only when the solvents have almost completely evaporated. A brief, but as high a contact pressure as possible is required for bonding. The strength of the adhesive joint is relatively high immediately after contact bonding and may amount to 50 % of the final strength. In principle, the contact adhesion effect may be obtained with natural rubber solutions, as known, for example, from the repair of bicycle inner tubes with rubber solution. However, the actual contact cements are solutions of synthetic rubbers with resins or solutions of high molecular mass polyurethane elastomers. The rubber components are mainly polychloroprene, nitrile – butadiene, and styrene – butadiene rubbers, and the resins used are primarily phenolic resins, rosins, and hydrocarbon resins. In the case of polychloroprene adhesives, metal oxides are added as stabilizers. In some cases, the rubber starting materials are high molecular mass rubbers that must be masticated on rolls or in internal mixers before dissolution, but mainly directly soluble rubbers are used today. The rubber, resin, and additives are dissolved in the appropriate solvents or solvent mixtures in closed stirring vessels to form the end product. The solid contents of the products varies from 20 to 30 %; the more sprayable types have lower solids contents.

The highest strengths are obtained with a uniform, thin coating of adhesive coupled with a high contact pressure. It is important to allow for adequate evaporation of the solvents and to observe the open time of the contact adhesive.

Contact cements based on nitrile rubber and polyurethane are more resistant to oils and plasticizers than polychloroprene adhesives. To improve adhesion and to increase thermal stability, contact adhesives may be applied together with isocyanate hardeners. Today solvent-free contact adhesives based on aqueous polychloroprene latexes are also available.

Trade Names. In Germany: Pattex, UHU Greenit; in Japan: Cemedine, Bond G, Sunstar; in the USA: Pliobond, Fastbond, Elmer's.

Pressure-sensitive adhesives are permanently tacky substances that adhere spontaneously to the surfaces of most materials with only light pressure [36]. Their adhesiveness is relatively nonspecific, although the individual products show different adhesion values on different substrates. Pressure-sensitive adhesives are used mostly for coating supports, for example, paper and film tapes, adhesive labels, or self-adhesive decorative sheeting. The adhesives are applied to the supports as solutions, dispersions, or melts. After the adhesive has set, the pressure-sensitive adhesive is present as a thin film on the support and, until it is used, is covered by suitable antiadhesively finished materials, for example silicone paper or siliconized polyethylene film. In some cases, the pressure-sensitive adhesive is also applied to the masking material, and the film of pressure-sensitive adhesive formed is transferred to the actual carrier by the reverse process. The dry application weight varies between 4 and 40 g/m^2, depending on the end use and the roughness of the carrier or substrate to be bonded.

In general, bonds formed with pressure-sensitive adhesives show relatively high resistance to brief and rapid loads. Under permanent load or stress, however, they tend to creep, particularly at elevated temperature. The property spectrum of pressure-sensitive adhesives is characterized by adhesiveness and resistance to heat, aging, and plasticizers.

The raw materials used for pressure-sensitive adhesives are natural and synthetic rubbers in conjunction with modified rosins, phenol–formaldehyde resins, or hydrocarbon resins. In addition to rubber, polyacrylates, polymethacrylates, poly(vinyl ethers), and polyisobutenes also are used frequently, again mostly in combination with resins. Silicone resins are used for special applications.

Pressure-sensitive adhesive dispersions are based primarily on special acrylic ester copolymers, again generally in combination with resins. In addition to suitable resins, two polymer bases are used primarily for hot-melt pressure-sensitive adhesives, namely, ethylene–vinyl–acetate copolymers and styrene–butadiene or styrene–isoprene block copolymers, which also are known as thermoplastic rubbers.

It is said that a pressure-sensitive adhesive always must be composed of a high-polymer base resin, which determines cohesion and specific adhesion, and of tackifying resins (tackifiers). In some systems, the tackifier may be replaced by low molecular mass fractions of the base polymer. To increase cohesion, the base resin in some systems is cross-linked or, in the case of rubber-based pressure-sensitive adhesives, vulcanized after application.

1.6.1.3. Adhesive Solutions from Which the Solvents Evaporate during Bonding

Solutions of Polymeric Substances and Resins in Organic Solvents. A distinction is be drawn between adhesives that set by evaporation of the solvent and form an adhesive joint by adhesion of the dissolved polymer to the substrate surfaces, and adhesives for plastics, the solvent of which dissolves and swells the plastics surfaces and joins them by means of migration processes. This allows bonding of plastics that are otherwise difficult to join. The polymer solutes in solution adhesives act primarily as thickeners for establishing certain flow properties and retaining the solvent for the duration of the dissolution process. These solid components also may perform a limited gap-filling function.

Adhesion Adhesives. The best known solvent-containing adhesives are the *multipurpose adhesives*. They can not join every type of substrate, but can bond a wide range of substrates with different results. They contain as binders high molecular mass compounds that adhere to many different surfaces, preferably nitrocellulose and poly(vinyl acetate). The solvent mixtures consist of esters and ketones and, in some cases, also contain alcohols. The composition of the solvent is determined by the polymers used, by the required fluidity of the adhesive, and by the required open time or setting time. The solvents may dissolve certain plastics.

Solutions of natural and synthetic rubber, polyurethane rubber, ethylene–vinyl acetate copolymers, polyacrylates, and other polymers with additions of resins and plasticizers are also used. These polymers can be used in solvents or as dispersions in water.

Trade Names. In Germany: UHU Alleskleber, Pritt Alleskleber; in Japan: Saivinol, Cevien-A; in the USA: Elmer's School Glue, Wilhold Glu-Bond.

Solvent Adhesives. In solvent adhesives for bonding plastics, the composition of the solvent is particularly important with regard to dissolution and swelling of the plastics to be bonded. Synthetic resins from the same class or from a related class are used for thickening.

The PVC adhesives have been used since the 1930s for bonding PVC resins. They are solutions of chlorinated poly(vinyl chloride) in acetone, cyclohexanone, THF, or mixtures thereof. The THF-containing adhesives are solutions of PVC in mixtures of THF, 2-butanone, and cyclohexanone. Organic thickeners are used to obtain pseudoplastic solutions that can be applied in sufficiently thick layers. By virtue of their stronger dissolving and swelling effect, THF adhesives can bridge certain gaps.

Pure solvents, such as aromatic hydrocarbons, halogenated hydrocarbons, solvent mixtures, or even solutions of polystyrene in these solvents are used for bonding polystyrene, ABS, and other styrene copolymers.

Trade Names. Solution adhesives for rigid PVC: Tangit, Dytex (Germany); Hamatide, S-dine (Japan); Bostik, Datey (USA).

Solutions of Natural and Synthetic Polymers in Water. *Aqueous Starch- and Dextrin-Based Adhesives.* The starch- and dextrin-based adhesives reached their zenith between 1920 and 1935. Since then, they have been increasingly superseded by synthetic resin-based adhesives, and they are now important only in certain applications. However, they are still used in considerable quantities for applications such as labeling adhesives, paper-sack adhesives, and envelope gums.

The raw materials used for the various types of starch- and dextrin-based adhesives are chiefly potato starch and cornstarch and, to a far lesser extent, tapioca starch and wheat starch. Potato starch is preferred for liquid glues, at least in Europe, where it is widely available.

Native *starch* is not soluble in water. To obtain soluble starches, the starch granule must be gelatinized by the prolonged heating of an aqueous starch suspension above the gelatinization temperature (e.g., 62.5 °C for potato and cornstarch, 67.5 °C for wheat starch) or by treatment with alkali at lower temperature. The gelatinization temperature and also the gelatinization rate can be varied by adding electrolytes.

To obtain certain solution viscosities and structures, the starch molecule must be degraded before and after gelatinization. A starch suspension is exposed before gelatinization to the action of acids, primarily nitric acid, or oxidizing agents, such as hydrogen peroxide, at elevated temperature. The greatest degradation is carried out in the case of dextrination. Because of the different degrees of degradation, the various gelatinization processes, and additives that influence hydration, the adhesives obtained differ considerably in their rheology, adhesive strength, and application properties; they are marketed as ready-to-use adhesives and as cold-water- or warm-water-soluble powdered adhesives. All aqueous starch- and dextrin-based adhesives must be protected against biological degradation by additions of preservatives [37].

Dextrin-based adhesives are available as powdered dextrins, which are dissolved in water before application, and ready-to-use aqueous dextrin adhesives with a dextrin content of 50 – 70 % and viscosities of 3×10^4 to 3×10^7 mPa s.

Dextrins are produced by the thermal and acid-hydrolytic degradation of native starch, generally cornstarch and potato starch. Dextrin-based adhesives are produced by dissolving powdered dextrins or dextrin mixtures of differing degrees of degradation in water. Mixtures of 60 – 70 % dextrin and 30 – 40 % water are heated in glass-lined vessels to 80 – 100°C until a homogeneous solution is formed. Commercial dextrin-based adhesives differ in the type of dextrins or dextrin mixtures used, the solids content, and in the additives present, such as hygroscopic salts ($CaCl_2$, $MgCl_2$), borax, glycerol, sugar or sugar syrup, and similar substances).

The viscosity of dextrin-based adhesives can be reduced by vigorous stirring. However, the viscosity of dextrin-based adhesives generally undergoes a marked increase after production, so that they have to be left to mature. In most cases, dextrin-based adhesives are marketed only after the maturing process is complete.

Mixed Glues. Starches and dextrins also are used in combination with other raw materials in adhesives. For example, the resistance to water of starch-based adhesives can be improved by the addition of melamine – formaldehyde or urea – formaldehyde resins. Mixtures of starch-based adhesives or dextrin-based adhesives with synthetic resin dispersions are also used.

Casein glues are also still important. Casein is obtained by precipitation from milk protein. The dried powder is mixed with water prior to application and hardens by loss of water at room temperature, with some chemical conversion of protein to more insoluble calcium derivatives. This adhesive is unsuitable for outdoor use but is more resistant to temperature changes and moisture than other water-base adhesives. Casein adhesives tolerate dry heat up to 70 °C and have good resistance to organic solvents. Although they are susceptible to biodegradation, chlorinated phenols can be used to inhibit this behavior. Typical applications include packaging (paper labels for glass), woodworking, and fabrication of interior-grade plywood.

Other Aqueous Adhesive Solutions. Besides starches and dextrins, *cellulose ethers,* primarily methyl cellulose and carboxymethyl cellulose, and in some cases also *starch ethers,* are important as wallpaper pastes and poster glues. The products are generally marketed in the form of powders that are dissolved in water by the user. Aqueous solutions of *poly(vinyl alcohol)* can be used to a limited extent. In most cases, fillers are added to these solutions.

Poly(vinyl pyrrolidone) has acquired significance as a binder in glue sticks. In addition to the polymer, these sticks consist primarily of a soap/water gel as builder. After application, the water evaporates, leaving the solid behind as bonding agent [39].

Aqueous Glutins (Animal Glues). The significance of animal glues or glutins has steadily diminished in the last decades. Like gelatin, glutines are obtained by acidic or alkaline hydrolysis of the collagen present in animal bones, hides, etc. The degradation product, which forms colloidal solution in water, is called glutin. Particularly pure glues are known as gelatin glues. In some cases, gelatin waste is also made up into glues.

The glutin solutions that accumulate in the processing of bones and hides are concentrated to a solids content of 30 – 55 %. They solidify on cooling to form firm, cuttable jellies. The slabs or blocks are further reduced in size, dried to a solids content of 85 %, and marketed in various forms. Hide glues are stronger than bone glues and often exceed the strength of wood. These adhesives are primarily used to bond wood, leather, paper, and textiles. Use is limited to interior applications. The setting of glutins is determined primarily by solidification of the gelatin or glutin/water mix on cooling and secondarily by evaporation of the water.

The reversible uptake and release of water by glutins has advantages and disadvantages. The possibility of activating layers of glue jellies by moistening is utilized, for example, in the production of moistenable gum tapes [40].

1.6.1.4. Aqueous Emulsions of Polymeric Compounds

Emulsion-based adhesives are adhesives that contain aqueous synthetic resin dispersions or emulsions or rubber latices as their principal raw material or active binder [41]. They are milky white, odorless liquids with solids contents of 40 – 60 %. In quantitative terms, emulsion-based adhesives are one of the most important groups of adhesives. They set by the release of water and the formation of a film from the emulsified or dispersed resin or rubber droplets.

The number of resin emulsions and rubber latexes suitable for use as raw materials for adhesives is very large. Emulsions of homopolymers and copolymers of vinyl acetate are the most important. The comonomers for vinyl acetate include maleic esters, acrylates, ethylene and vinyl chloride, unsaturated carboxylic acids, and vinyl esters of relatively long-chain fatty acids, such as vinyl laurate. Apart from these products, polyacrylate homopolymers and copolymers and also styrene copolymers are particularly important in adhesives [42].

Additions of plasticizers, solvents, and resins improve the specific adhesion of the synthetic resin emulsions, and also are capable of lowering the film-forming temperature and increasing the tackiness of the wet adhesive. Poly(vinyl alcohol), cellulose ethers, and other hydrocolloid additives prolong the open time. Poly(vinyl alcohol) also can improve wet tack. Certain applications require fillers, which may be used in such large quantities that pasty mortars and cements are formed.

The elasticity of the set films can be increased by plasticizers and by copolymerization. Redispersion powders, that is, spray-dried synthetic resin emulsions, are added to synthetic resin emulsions to increase their solids content and their setting rate.

Synthetic resin emulsions containing reactive groups also have appeared on the market. The reactive groups provide for subsequent cross-linking, making it possible, for example, to increase the resistance of the joints to water. Mixed adhesives consisting of synthetic resin emulsions and starch- or dextrin-based adhesives have advantages in certain fields and are generally cheaper.

Latex adhesives are based on natural rubber or on various types of synthetic rubber. Occasionally, mixtures of natural rubber and synthetic resin latices are also used. In general, resins or solvents also are used in the production of latex adhesives.

In the production of latex adhesives, overvigorous stirring can affect the rheological properties of the adhesives, which are often pseudoplastic and differ in their mechanical stability. Characteristic properties of emulsion-based adhesives include viscosity and flow properties, the setting and breaking behavior of the emulsion, the wet tack or green strength of the film, the minimum film-forming temperature, storage behavior at low temperatures (freezing temperature), and specific adhesion to various materials. Films of emulsion-based adhesives are tested only occasionally for tensile strength and elasticity. The redispersibility of the set adhesive films is important with regard to the cleaning of machine components and recycling of paper.

1.6.2. Adhesives Setting by Chemical Reaction

Before setting, reactive adhesives consist predominantly of reactive low molecular mass monomers and/or oligomers that, during curing, are converted by chemical reactions into high molecular mass, often three-dimensionally cross-linked polymers. Reactive adhesives can be divided into polymerization, polyaddition, and polycondensation adhesives. Vulcanizing reactive adhesives are a special group.

Reactive adhesives are marketed as one-pack, two-pack, and no-mix formulations. Two-pack adhesives must be completely mixed in the correct ratio before use. The mixtures have a limited pot life. One-pack adhesives are easier to handle during application. They are hardened by heat, by the catalytic action of the substrate, or by atmospheric moisture.

Cold-setting reactive adhesives harden within a few hours to several days at room temperature or slightly elevated temperature. An exception are cyanoacrylate adhesives, which cure on several substrates within seconds or minutes, as do UV/EB curable adhesives. Warm-setting adhesives generally require temperatures in the range 80 – 100 °C, and hot-setting adhesives 100 – 250 °C. Adhesive joints of particularly high quality are obtained mainly with warm- and hot-setting systems. In many cases, warm setting leads to higher bond strengths, even in the case of systems that normally harden at room temperature. However, even certain cold-setting adhesives produce outstanding bond strengths under favorable bonding conditions.

To avoid mixing errors in the case of two-pack adhesives, no-mix adhesives are available. In this case, one of the surfaces to be bonded is pretreated with the adhesive resin, the other with the adhesive primer, and the bond is formed instantly when the two surfaces are brought together.

1.6.2.1. Adhesives Setting by Polymerization

Polymerization adhesives harden through radical or ionic polymerization of the monomers; at the same time, graft polymerization or cross-linking of the dissolved, still polymerizable polymers also may occur.

Two-Component Polymerization Adhesives. The best known types are polymerization adhesives based on solutions of *unsaturated polyesters* in styrene or (meth)acrylates. Peroxides are added as hardeners to the resin component; amines or heavy-metal salts are used as accelerators.

More important and with better properties are the cold-setting, two-component *acrylate adhesives,* which contain methacrylates or acrylates, sometimes mixed with styrene and methacrylic acid as monomer and, in addition, various polymers. The polymers used are primarily synthetic rubbers, such as polychloroprene, styrene – butadiene rubber, butyl rubber, polystyrene, polymethacrylates, and acrylate graft polymers of these polymers. Amines are used as accelerators, and benzoyl peroxide in the form of plasticizer pastes or a powder mixture with fillers is preferred as hardener [43].

Trade Names. In Germany: Agomet, Pattex Stabilit Express; in Japan: Hard-Lock, Sumikitt; in the USA: Versilok.

One-Component Polymerization Adhesives. *Cyanoacrylate Adhesives.* Methyl, ethyl, butyl, and methoxyethyl esters of cyanoacrylic acid are used for cyanoacrylate adhesives [44]; soluble polymers and plasticizers are incorporated to regulate viscosity and for elastification. Cyanoacrylate adhesives rapidly polymerize by an ionic reaction mechanism initiated simply by weak bases to form high molecular mass, but largely uncross-linked polymers. In most cases, atmospheric moisture or the film of moisture on the substrate is sufficient to initiate polymerization, because the adhesives are applied in very thin layers. However, this sensitivity to atmospheric moisture means that the adhesives must be stored in tightly sealed form, usually in polyethylene bottles. Cyanoacrylates are used for bonding small items of nearly all substrates. In case of polyethylene or polypropylene special primers are available. Setting is complete with rubbers in seconds, with aluminum in less than one minute.

Cyanoacrylate adhesives are also used in surgery [45]. They enable parts of the body to be joined together and, therefore, must be used carefully.

Trade Names. In Germany: Sicomet, Loctite, Pattex Blitzkleber; in Japan: Aron Alpha, Alteco, Cianobond; in the USA: Superbonder, Permabond, Super Glue, Krazy Glue and Attak.

Anaerobic adhesives are mixtures of methacrylic esters which remain liquid when exposed to air but harden when confined between metal surfaces. These mixtures can be used for a large number of industrial purposes such as locking threaded fasteners, sealing threaded pipe connections, retaining cylindrical machine components, sealing flange joints, bonding structural components, sealing porous metal castings, welds and powdered metal parts and many other applications which are still being found more than forty years after the initial invention [134]. Several reviews have been published which describe anaerobic adhesives and their applications [135]–[137].

All anaerobic-curing adhesives consist of a monomer, an initiator, one or more accelerators, and stabilizers. The properties of the cured and uncured adhesives can be modified to control viscosity, color, or fluorescence, reduce strength, increase toughness or heat resistance, provide lubrication and reduce settling of solid fillers. In some applications anaerobic sealants cure more rapidly if the surface is treated with chemicals which catalyze their polymerization.

A number of papers discuss the *reaction mechanisms* of anaerobic adhesive cure [138]–[142]. The polymerization mechanism of anaerobic adhesives is similar to that of other radical initiation systems except for the special way in which the inhibiting effect of oxygen is used to delay the polymerization and in the chemical activation which occurs at the metal surface.

Initiation:

$$I \longrightarrow I^{\cdot}$$

$$I^{\cdot} + M^{\cdot}$$

Inhibition:

$$M^{\cdot} + O_2 MOO^{\cdot}$$

Propagation:

$$M + M^{\cdot} \longrightarrow M^{\cdot}$$

Termination:

$$M^{\cdot} + M^{\cdot} \longrightarrow M_2$$

$$M^{\cdot} + I^{\cdot} \longrightarrow MI$$

The reaction rate of oxygen with free radicals is very high, and the resulting peroxy radical is a relatively poor initiator. When the supply of oxygen is used up within a thin bond line the propagation step can provide rapid development of adhesive strength.

An important factor in the initiation of anaerobic adhesive cure is the redox reaction between a hydroperoxide and transition metals with adjacent oxidation states [138].

$$Fe^{2+} + ROOH \longrightarrow Fe^{3+} + RO^{\cdot} + OH^{-}$$

$$Fe^{3+} + ROOH \longrightarrow Fe^{2+} + ROO^{\cdot} + H^{+}$$

Other transition metals react similarly, and copper is particularly active.

The use of saccharin and *N,N*-dimethyl-*p*-toluidine results in a substantial acceleration of the initiating reaction. Although each of these components is an accelerator by itself the combination has a strong synergistic effect. It has been suggested that a charge-transfer complex is formed by these materials [139]. It has been suggested that one of the functions of saccharin in anaerobic curing is to dissolve metal ions from the surface that then catalyze the decomposition of CHP [142]. The role of the bonding surfaces and the effect of different types of accelerators account for most of the literature on the cure mechanism of anaerobic adhesives.

The first patents on anaerobic adhesives mentioned only the polyglycol dimethacrylates as *monomers*, mainly tetraethyleneglycol dimethacrylate [143], [144]. The use of acrylic or methacrylic acids to improve adhesion to smooth surfaces was mentioned in [145]. A series of polyurethane polyacrylates were prepared by reaction of toluene diisocyanate, other isocyanates, and isocyanate-terminated oligomers with hydroxyalkyl methacrylates [146]. These monomers could be tailored to provide the strength and toughness for some structural adhesive applications. The incorporation of hard and soft segments into the polyurethane backbones provided significant improvements in toughness, cure-through-gap, and cryogenic strength properties [147].

The use of monomethacrylates in anaerobic formulations was disclosed in a patent assigned to Loctite. Specifically mentioned were hydroxyethyl, hydroxypropyl, cyclohexyl, tetrahydrofurfuryl, dimethylaminoethyl, and glycidyl methacrylates and cyanoethyl acrylate [148].

Methacrylate esters containing residual carboxylic acid groups were prepared by the reaction of hydroxyethylmethacrylate with phthalic anhydride, pyromellitic dianhydride, and benzophenone tetracarboxylic acid dianhydride. The residual acid provided improved adhesion.[149], [150] The reaction product of hydroxyalkylmethacrylates with maleic anhydride also produced monomers with residual acid as well as additional curable unsaturation [151].

Three Bond Company used trimethylolpropane trimethacrylate [152] and ethoxylated bis-phenol A dimethacrylate [153] in anaerobic formulations. These monomers have some advantages in providing improved heat resistance.

At Henkel dicyclopentadienyl methacrylate was used in anaerobic formulations with high strength [154]. Rohm & Haas disclosed the use of dicyclopentenyloxyethyl acrylate and methacrylate in anaerobic formulations [155]. These monomers provide good cure strength on metal parts which have not been degreased and also have lower odor and volatility than the corresponding dicyclopentadienyl esters.

Silicone methacrylates have been formulated into anaerobic adhesives by Dow Corning [156], Toshiba Silicones [157], and Loctite [158], [159].

The most commonly used *initiator* for anaerobic adhesives is cumene hydroperoxide (CHP). Many other hydroperoxides have been disclosed such as *tert*-butyl hydroperoxide, *p*-menthane hydroperoxide, diisopropylbenzene hydroperoxide, pinene hydroperoxide and methyl ethyl ketone hydroperoxide [160].

Storage-stable anaerobic formulations can be prepared without hydroperoxide if the methacrylate resin is aerated in the presence of an amide and a tertiary amine [161].

The first *accelerator* used in an anaerobic adhesive was tri-*n*-butylamine.[162] Saccharin was also found to be an effective accelerator [163] and the combination of saccharin and *N,N*-dimethyl-*p*-toluidine was particularly effective if properly stabilized [164].

Many patents have been issued on various accelerators and combinations thereof. 1-Acetyl-2-phenylhydrazine [165], [166], benzenesulfonyl hydrazide [167], dibenzenesulfonamide [168], and other similar compounds have been disclosed.

The use of saccharin is of particular interest, and a number of compounds have been prepared which have a similar chemical structure and reactivity. The reaction of sulfuryldiisocyanate with acetic acid gives a disulfonamide which is an effective accelerator [169]. Similar reactions of *p*-toluenesulfonyl isocyanate and chlorosulfonyl isocyanate can be used to prepare many different compounds which are active accelerators [170], [171]. These methods allow the preparation of accelerators with improved solubility.

The monomers used in anaerobic adhesives and sealants generally contain at least one free radical *stabilizer* such as hydroquinone or *p*-methoxyphenol. Benzoquinone, naphthoquinone, and similar compounds provide improved shelf stability without retarding the anaerobic cure [172].

The use of a soluble metal-chelating agent such as tetrasodium ethylenediamine tetraacetate effectively stabilizes an anaerobic formulation against small amounts of metal contamination [173]. The wide variety of applications of anaerobic adhesives and sealants is made possible by the modifications which make the viscosity appropriate to the application. An application which requires penetration into close fitting parts should have very low viscosity while a product used with large, loose fitting parts should have a high viscosity. Polymethacrylates, cellulose esters, butadiene – styrene copolymers, acrylonitrile – butadiene – styrene copolymers, poly(vinyl chloride), copolymers of vinyl chloride and vinyl acetate, poly(vinyl acetate), cellulose ethers, polyesters, polyurethanes, and other thermoplastic resins have been used to control the flow characteristics of anaerobic sealants [174].

The flow characteristics of anaerobic formulations can also be controlled by the addition of fumed silica and other solid additives which can impart thixotropic properties [175].

Many anaerobic adhesives and sealants may be required to develop a relatively low strength so that the components can be disassembled for repair or replacement. Many liquid plasticizers have been used for this purpose [176], but the use of a low molecular mass polyester is advantageous [177].

The "toughness" properties of anaerobic adhesives can be enhanced by the addition of a reactive elastomer [178]. The heat resistance of anaerobic adhesives and sealants can be enhanced by the addition of a bis-maleimide [179]. These maleimide additives appear to be relatively unreactive during the initial anaerobic cure. As the adhesive is exposed to high temperatures, the methacrylate backbone degrades and the methacrylates can then copolymerize with the maleimides to form a more heat resistant matrix [180].

The addition of dyes to anaerobic adhesives and sealants assists in identification and inspection of the products. Automated inspection procedures are made possible with dyes that fluoresce under ultraviolet light. Titanium dioxide pigments can make the sealants opaque and more visible.

Solid fillers are added to some anaerobic adhesives and sealants for various purposes. Mica, talc and other mineral fillers can help to provide an "instant seal" capability to anaerobic pipe sealants. The sensitivity of the anaerobic cure system to metal contamination requires that these fillers be chosen very carefully.

Powdered graphite, polytetrafluoroethylene, and polyethylene can function as lubricants in pipe sealants and threadlocking compounds. This lubrication can prevent galling in close-fitting pipe threads [181]. Lubricating additives in thread-locking sealants can provide control of the clamping force exerted by a fastener at a given tightening torque.

The cure speed and adhesion of an anaerobic sealant can be increased by treatment of the surface with a solution of a *primer* or *activator.* Early anaerobic thread lockers were strongly affected by part cleanliness, and degreasing the parts with a chlorinated solvent improved performance dramatically. The condensation product of an aldehyde and a primary or secondary amine, a sulfur-containing radical accelerator, or a compound of an oxidizable transition metal were some of the materials used to activate the anaerobic cure [182] – [184]. Due to the ozone-depleting potential of chlorinated solvents, acetone, alcohols, or hydrocarbon solvents are generally used at present.

Modified acrylics, also referred to as second-generation acrylics or no-mix adhesives, are composed of a modified acrylic adhesive and a surface activator. Typical modified acrylics are based on cross-linked polymethyl methacrylate grafted to vinyl-terminated nitrile rubber. Carboxyl-terminated rubbers have also been used.

Unlike epoxies, which cure by an ionic polymerization mechanism, modified acrylics cure by free-radical addition. Therefore, careful proportioning of components is not required. In two-component systems, no mixing is required because the adhesive is applied to one substrate, the activator to the other, and the substrates are joined. Handling strength is rapidly achieved with this fast-curing system.

Modified acrylics have good peel, impact, and tensile lap-shear strengths. High bond strengths are obtained with metals and plastics even if surfaces are oily or improperly cleaned. The cured adhesive exhibits little shrinkage. Resistance to high humidity is good, particularly when bonding plastic substrates. The limitations of modified acrylic adhesives include low high-temperature strength, flammability of the primer, which is an amine (Schiff base) or a peroxide in a solvent, and the odor of the uncured adhesive.

Trade Names. In Germany: Agomet, Multi-Bond; in Japan: Diabond SGA, Cemedine SGA; in the USA: Versilok.

A/B acrylates are a new class of acrylate adhesives. They are two-component adhesives that cure rapidly after application. They avoid using a solvent-based primer as in modified acrylics. The acrylates used are mainly acrylated polyurethane oligomers, compounded with catalyst in one component and accelerator in the other. The two components can be applied 1/1 without mixing as beads side by side. Mixing is performed by coating the materials to be bonded or by a static-mixer applicator. Curing starts immediately after mixing. Handling strength is attained rapidly, and final curing is complete after several hours. An advantage over modified acrylics is the lesser odor due to the use of oligomers instead of monomers.

Trade Names. In Germany: Omnifit A/B; in the USA: Loctite.

1.6.2.2. Adhesives Setting by Polyaddition

Epoxy resin Adhesives. Epoxies [47], [48] are among the most widely used structural adhesives. These chemically reactive systems include two-component systems that cure at room temperature or elevated temperature and one-component systems, which usually require heat for curing.

Epoxy thermoset adhesives, available in many formulations, can be used to join most materials. These adhesives have good strength, do not emit volatile substances during curing, and have low shrinkage. However, peel strength and flexibility are low, and epoxies are brittle.

A typical two-component system consists of a resin and a hardener, which are packaged separately. In a one-component system, the resin and hardener are packaged together. Other possible additives include accelerators, reactive diluents, plasticizers, resin modifiers, and fillers.

The most common epoxy resin is based on the diglycidyl ether of bisphenol A. Epoxy resins are sometimes esterified with fatty acids or modified with polyurethane intermediate chains for elastification.

The hardeners used for *hot-setting formulations,* i.e., for hardening temperatures above 100°C, are preferably dicarboxylic acid anhydrides, dicyandiamide, and certain aromatic amines. The hardening temperatures, which are normally in the range 150 – 200°C, can be lowered by adding accelerants [49].

The hardeners for *cold-setting systems* include aliphatic and cycloaliphatic amines and polyamines, adducts of polyamines and epoxy resins, phenol – amine combinations, and condensates of polyamines and dimerized fatty acids (polyaminoamides). Whereas amine hardeners must be used in a stoichiometric ratio to the reactive epoxy groups, polyaminoamides may be overdosed to a certain extent and thus used to elasticize the adhesive resin.

An exothermic reaction occurs during the curing of epoxies. This reaction can be minimized by lowering the temperature of the mixed components in two-component systems, by limiting the batch size, and by using shallow mixing containers.

Epoxies are brittle, especially if cured with an anhydride. Therefore, thermoplastics and rubbery modifiers are often added to decrease the brittleness.

Nylon epoxy, which first became available in 1960, has an approximate maximum service temperature of 180 °C, compared to 80 °C for unmodified epoxies. The main advantage of nylon epoxies is increased flexibility and much higher peel strength, compared to unmodified epoxies. Also this very tough material has excellent tensile lap-shear strength. Fatigue and impact resistance are also good. Limitations include poor peel strength at low temperatures and poor creep resistance. Perhaps the most serious limitation is extremely poor moisture resistance in both the uncured and cured material. Nylon epoxies are not as durable as elastomer epoxies or thermoplastic-modified epoxies. A typical application of nylon epoxies is in laminates.

Elastomer epoxies generally contain nitrile rubber as the elastomeric component. This system is also referred to as a modified or toughened epoxy. One of the applications of widest use is in films and tapes. Elastomer epoxies cure at low pressures and low temperatures over a short time interval. This is achieved by adding a catalyst to the adhesive formulation. Bond strengths of elastomer epoxies are lower than those of nylon epoxies. However, the major advantage of elastomer epoxies is their sub-zero peel strengths, which do not decrease as fast as those of nylon epoxies. In addition, the moisture resistance of elastomer epoxies is better than that of nylon epoxies but not as good as that of vinyl – phenolics or nitrile – phenolics. Limitations to the use of elastomer epoxies include poor water immersion resistance and poor properties when exposed to marine conditions.

In addition to liquid and pastelike epoxy resin adhesives, solid powdered products are also available commercially. In addition, epoxy resins are used in the form of adhesive films for certain applications. The most important are nylon epoxy resin films consisting of mixtures of relatively high molecular mass polyepoxides and polyamides, and epoxy resin – phenolic resin combinations.

Trade Names. In Germany: Araldite, Metallon, Pattex Kraft-Mix, UHU-plus, Scotch weld, Terokal; in Japan: Sho-bond, KBK, Bond E, Sanyu Resin; in the USA: Scotch Weld, Resiweld, Ross Epoxy Glue, Ten-set Epoxy and Duro E-pox-e Glue.

Reactive Polyurethane Adhesives date back to the late 1930s and acquired their present economic significance over the past decades [50] – [52]. They are distinguished by good adhesion to various substrates and, compared with other reactive adhesives, by high elasticity, even at low temperatures. Polyurethane adhesives are marketed both as solvent-free and as solvent-based adhesives.

The generic term polyurethane adhesives covers both adhesives that already contain polyurethane compounds or compounds with a urethane bond and also reactive adhesives in which urethane groups are formed only during the hardening process.

Chemically reactive polyurethanes include both one- and two-component systems. One-component systems are usually based on a polyether polyol treated with a polyisocyanate to give an isocyanate-terminated polymer. A one-component system cures when exposed to moisture at room temperature. One-component polyurethane hotmelt adhesives are also cured by moisture after application. Two-component systems result from the reaction of low molecular mass polyols and isocyanates or from isocyanate-terminated prepolymers with either polyols or polyamines. Two-component systems cure at room and/or elevated temperatures.

One-component heat-curable urethanes are also available. In these formulations, free isocyanate groups are typically blocked by the addition of phenol. The prepolymer is then blended with the polyglycol curing agent and packaged. The mixture is stable until heated to elevated temperatures, whereupon the phenol is released and the isocyanate groups are regenerated. A rapid cure occurs when these groups come into contact with the polyglycol.

Polyurethanes bond to most surfaces. They also have outstanding tensile lap-shear strength at lower temperatures. In fact, polyurethanes have better low-temperature strength than any other adhesive, even epoxy – nylons. Good flexibility, abrasion resistance, and toughness are other advantages of polyurethanes. Limitations include sensitivity to moisture in both uncured and cured adhesives, the toxicity of isocyanates, and poor tensile lap-shear strength at higher temperature. A major advantage, and also the reason for the wide range of applications of polyurethane adhesives, lies in their versatility in regard to raw materials, composition, and curing.

Normally, reactive polyurethane adhesives contain high molecular mass, still reactive polyurethane prepolymers with terminal hydroxyl or terminal isocyanate groups. Prepolymers with terminal hydroxyl groups can be hardened by the addition of polyisocyanates. Depending on the functionality of the prepolymers and polyisocyanate hardeners, curing may involve both cross-linking and linear enlargement of the molecule. High molecular mass polyurethane elastomers contain urethane groups that can still react with isocyanates and are thus cross-linked (contact cements).

Prepolymers with terminal isocyanate groups can react with and are hardened both with amino or hydroxyl compounds and with water. Hardening with water is preferred for one-pack systems, for which both atmospheric moisture and the film of water on the substrates act as hardeners for the prepolymer. Cross-linking with polyamines is particularly fast compared with curing by polyols.

One-pack polyurethane adhesives can also be manufactured by protecting the isocyanate groups of prepolymers with, for example, phenols or oximes, which are split off by the action of heat during bonding. Protected polyisocyanates can also be added to reactive polyurethane adhesives as latent hardeners. Tertiary amines or organotin compounds can be added to accelerate isocyanate reactions.

Polyurethane adhesives should be stored in tightly sealed containers, because shelf life and adhesive quality can be affected adversely by the uptake of water.

Trade Names. In Germany: Liofol, Macroplast; in Japan: Diabond, Bostik; in the USA: Chemlok.

1.6.2.3. Adhesives Setting by Polycondensation

Polyhydroxymethyl Compounds. Hydroxymethyl compounds that release water during condensation are used primarily for polycondensation adhesives. Accordingly, the high-quality bonding of impermeable substrates requires relatively high pressures to compensate for the water vapor pressure developed during hot setting. Polycondensation adhesives are used both in pure form and as solutions in water or organic solvents. Hardening can be initiated by the application of heat or by the addition of hardeners.

Polycondensation adhesives include phenolic and resorcinol resins and also urea – and melamine – formaldehyde resins. Adhesives of particularly high quality are obtained by combining phenolic resins with poly(vinyl formal) resins, nitrile rubber, or epoxy resins. Combinations such as these are often marketed as adhesive film.

Trade Names. In Germany: Redux, Tegofilm, Kaurit, Pressal.

Phenolics or phenol – formaldehyde structural adhesives are chemically reactive systems that cure to form thermosets. In one-component systems, meltable powders (resols) are used as binders for particle board or as alloys (including nitrile – phenolics, vinyl – phenolics, and epoxy – phenolics), which are used in the structural bonding of metals. In two-component systems, the resin and catalyst are mixed and then heated to initiate curing. Both systems cure by a condensation reaction that produces a byproduct.

In general, phenolics are low-cost adhesives with good strength and resistance to biodegradation, hot water, and weathering. Elevated-temperature resistance is also good. Limitations include low impact strength and high shrinkage stresses, which lead to brittleness. Shelf life is limited, the adhesives are dark colored, and they can be corrosive.

Phenolics dominate the wood adhesives market, especially for plywood. The structural adhesive bonding of metals, particularly with phenolic alloys, is another application.

Nitrile – phenolic alloys are composed of nitrile rubber and phenolic with additives. These systems are available in liquid or film form. The advantages include a maximum service temperature of 140°C, low cost, high bond strength, and excellent resistance to water, oil, biodegradation, and salt. Disadvantages include poor to moderately poor low-temperature resistance and high-pressure, high-temperature curing. Typical applications include the bonding of metals, plastics, rubber, wood, glass, and ceramics. This structural adhesive is used in automobile brake-shoes and clutch disks and in aircraft applications.

Vinyl – phenolics are alloys composed of polyvinyl formal (PVF) – phenolics or polyvinyl butyral (PVB) – phenolics. They are available as liquids and films. In general, vinyl – phenolics have a maximum service temperature of 80°C and are equal to phenolic – nitriles in strength. These adhesives are better than epoxies in sandwich structures that require high strength.

PVF – phenolics have good resistance to fatigue, weathering, fungi, salt, humidity, water, and oil. Creep resistance is good in some formulations, but is poor in others above 90°C. These adhesives are used in metal-honeycomb and wood – metal applications.

PVB – phenolics also have good resistance to weathering, fungi, salt, humidity, water, and oil, and creep and fatigue resistance are even better than those of PVF – phenolics. These adhesives are used to bond metal or reinforced plastic facings to paper honeycomb, cork to rubber, and steel to rubber.

Epoxy – phenolic adhesives are also available as liquids or films. They are among the best adhesives for long-term use at 150 – 250 °C. In general, epoxy – phenolics are relatively expensive. They have fairly good shear and tensile strengths over a wide temperature range, but peel strength is poor. Impact resistance is also poor, whereas resistance to weathering, aging, aromatic fuels, glycols, hydrocarbon solvents, and water is good. Typical applications include the bonding of metals, glass, ceramics, phenolics, and honeycomb sandwich structures in aircraft applications.

Silicone Adhesives. Cold-cross-linking one-pack silicone adhesives also harden by a polycondensation mechanism. They are silanols whose OH groups have reacted with acetic acid or amines. Treatment of these compounds with water leads to the elimination of acetic acid or amines and generates the reactive silanols, which cross-link with elimination of water. These products are particularly important as sealing compounds in the building industry.

Usually silicones are available as both one- and two-component systems that cure to thermoset solids. A one-component room-temperature-vulcanizing silicone cures at room temperature upon exposure to atmospheric moisture. Curing is either acidic or nonacidic in the presence of moisture. Adhesives that have an acidic cure have greater unprimed adhesion and a longer shelf life. However, the corrosion of metals by the acid is a potential problem. Thin films of approximately 0.6 mm cure within 90 min., whereas 15 mm films of require ca. 7 d to achieve full cure.

Two-component silicones do not require moisture to cure. During the addition polymerization of silicones, curing is achieved by catalytic action. Pot life, setting time, and cure time are dependent on the catalyst concentration. A system with 5 % catalyst will typically have a 3 h pot life, a 22 h setting time, and a 7 d cure time at room temperature. Increasing the temperature accelerates curing. Addition-polymerized silicones exhibit little shrinkage upon curing and have good high-temperature resistance.

In the condensation polymerization of two-component silicones, byproducts are released. These systems are less likely to be inhibited and can be used on a greater variety of substrates. However, reversion of polymerization is a potential problem.

In general, silicone adhesives have good peel strength over a service temperature range of – 60 to 250 °C. Some have limited service to 370°C. Flexibility and impact resistance are good, as are resistance to moisture, hot water, oxidation, and weathering. Typical lap-shear strengths are low in metal-to-metal bonds, with values ranging between 1 and 3.5 MPa. The cost of these adhesives is high.

Silicone adhesives are useful in bonding metals, glass, paper, plastics, and rubbers, including silicone, butyl rubber, and fluoroelastomers.

MS Polymers. A relatively new class of adhesives are the two-component MS polymer adhesives, known as moisture-cross-linkable sealants up to know. Due to their excellent elastic properties they are used for elastic bonding. Chemically they are based on silanealkoxy-modified prepolymers such as poly(propylene glycol). The end groups of these so-called MS polymers are silanealkoxyl groups which are hydrolyzed by water and then cross-link. In comparison to silicone adhesives, they adhere to all kinds of surfaces without any primer and are paintable in the freshly cured state.

Polyimides and Polybenzimidazoles [53]. The highly temperature resistant polyaromatic adhesive resins of the polyimide and polybenzimidazole type are applied as precondensates in the form of solutions or films and are condensed to completion at 230 – 250°C under a relatively high pressure of 8 – 10 bar. The precondensates remain stable for only a few hours at room temperature and have to be stored at –18 °C.

These structural adhesives with high-temperature resistance are based on synthetic organics with aromatic (benzene) and/or heterocyclic rings in the main structure. These chemical groups often include imidazoles and substituted imidazoles. These prepolymers have open-ring structures that close upon application of heat. The condensation reaction leads to a highly cross-linked structure. The adhesives are available as liquids and films. High-temperature adhesives, including polyimides and polybenzimidazoles, are expensive materials that are difficult to handle and require long curing times during which volatile substances are released. Polyimides are superior with regard to long-term strength retention at elevated temperatures. At 260 °C in air, polyimides have higher bond strength than epoxy – phenolics. Polybenzimidazoles are stable in air for short-term exposures up to 290 °C. Both polyimides and polybenzimidazoles are moisture-sensitive. Applications of high-temperature adhesives are primarily in the aircraft and aerospace industries for bonding metals.

1.6.2.4. Vulcanizing Adhesives

Adhesives for joining unvulcanized elastomers to metals or plastics under vulcanization conditions are called rubber-to-metal bonding agents. Mixtures of film-forming components, cross-linking agents, and stabilizers dissolved in organic solvents are used as bonding agents. The solids content varies from about 15 to 25 % according to type. The film-forming components used are chiefly halogenated polymers, whereas the cross-linking agents are, for example, nitroso compounds, isocyanates, and oximes in combination with an oxidizing agent [54]. In many cases, primers are applied additionally to the metal surface to improve resistance to corrosion.

Trade Names. In Germany: Chemosil; in Japan: Metalok; in the USA: Chemlok.

1.6.2.5. Ultraviolet/Electron Beam (UV/EB) Curing Adhesives

Most UV/EB adhesives are based on an addition polymerization curing mechanism. They consist of acrylic esters of various types or combinations of acrylates with aliphatic or aromatic epoxy resins, urethanes, polyesters, or polyethers. Although the epoxy-based systems have higher tensile strengths, their elongations are less than those of the urethane-based systems. In addition, the urethane-based systems have better abrasion resistance. UV/EB adhesives that undergo cationic polymerization are based on epoxies with reactive diluents and cyclic monomers. The main advantage of UV/EB-curable adhesives is rapid curing at room temperature, so they can be used to bond heat-sensitive substrates, such as poly(vinyl chloride). In addition, the rapid cure often eliminates the need to fix parts and greatly increases production rates.

UV/EB-cured adhesives have been used to replace solvent-base adhesives because of the increasing cost of properly recovering and disposing of solvents. Most of these adhesives are single-component materials that require no mixing and produce little waste.

The cross-linked nature of UV/EB-cured adhesives results in good chemical, heat, and abrasion resistance; toughness; dimensional stability; and adhesion to many substrates. Unlike thermal curing, EB or UV curing requires pure raw material grades. The use of dual-curing systems allows opaque substrates to be cured.

The mechanical properties of UV-curable structural adhesives are dependent on polymer molecular mass and cross-linking density. These factors are related to the prepolymer, degree of stiffness or flexibility, and functionality. The overall adhesive strength is affected by:

- Adequate UV transmission through the bond line
- Adhesive thickness
- UV intensity
- Postcuring (by heat or exclusion of oxygen, as in anaerobics)

The main components of UV/EB-cured adhesives are reactive oligomers which contribute adhesion, toughness, and flexibility to the overall properties of UV/EB adhesives. Typical reactive oligomers include acrylated epoxy resins and aromatic urethanes.

Monomeric diluents are low molecular mass monofunctional molecules that reduce the viscosity of liquid oligomers. Some, such as methacrylates, increase the toughness and adhesion. Various types of acrylates are used as reactive monomers.

Cross-linking monomers such as 1,3-butylene glycol dimethacrylate, tripropylene glycol diacrylate, and pentaerythritol tetracrylate are also used in UV/EB adhesive formulations.

Free-radical initiators trigger the cross-linking reaction. In EB-cured adhesives, the electrons act as free-radical initiators for addition polymerization. Therefore, no chemical initiator additives are needed. In UV-cured adhesives, photoinitiators, which release free radicals when exposed to UV radiation, are required to initiate addition polymerization. The most recent UV- and EB-curing systems involve cationic polymerization mechanisms.

Typical UV-curable adhesive applications include the electronics, automotive, medical, optics, and packaging industries, as well as tapes and labels. EB-curable adhesives are used in magnetic tapes and floppy disks, where magnetic particles are bonded to films, as well as in packaging, tapes, and labels.

Laminating adhesives have the largest share of the UV/EB-curable adhesives market. Adhesives are applied and then cured in-line. The lack of solvent in these adhesives eliminates the drying step required when using solvent-based adhesives. The high degree of cross-linking in UV/EB-cured laminates gives high bond strength, good heat resistance, and good chemical resistance. Substrates that are typically laminated include: film to paper, foil, fabric, glass, film, or wood; paper to foil, wood, or paper; wood to wood; and glass to glass or metal.

The pressure-sensitive sector of the UV/EB-curable adhesives market is not nearly as large as that of the laminating-adhesives segment. Because slight changes in exposure time greatly affect peel strength and tack, the formulation of these adhesives is critical. Currently, there are three approaches used to make pressure-sensitive adhesives. The first is based on a UV-curable system employing conventional tackifiers blended with UV-reactive moieties. The process must be closely controlled to attain repeatable tack and peel strength. The second approach is based on an EB-curable system that is actually a hot-melt adhesive. EB-cross-linkable thermoplastic rubbers are usually used in these adhesives and result in hot-melt pressure-sensitive adhesives with better heat resistance than conventional elastomeric hot-melt adhesives. The third approach involves the development of inherently tacky oligomers for pressure-sensitive adhesives.

There are several limitations to the use of UV/EB-cured adhesives. EB equipment is expensive. UV equipment is less expensive, but the materials themselves are usually more costly because of the need for photoinitiators. To cure adhesives properly, one substrate must be transparent to UV radiation. However, the necessity of having a transparent substrate has been removed by the introduction of dual-curing adhesives.

These adhesives are quickly set by a UV cure and are more fully cured by a second mechanism involving the introduction of heat or moisture or the elimination of oxygen (anaerobics). In EB-curable adhesive, the depth of EB penetration is limited by the density of the material, rather than its opacity.

The cure time of UV adhesives is usually less than 60 s and depends on:

- Bond-line joint thickness: As the thickness increases, UV radiation loses its ability to penetrate totally, necessitating a second cure.
- Type of substrate: A transparent substrate such as glass with a small gap may take as little as 5 s to cure. Opaque and darker substrates require longer curing times.
- Light intensity: The more intense the UV light, the faster the cure.

The curing time of EB adhesives is comparable to that of UV adhesives. In general, EB radiation allows adhesive curing to be achieved at greater depths than is possible with UV radiation. Electrons can pass through substrates that are opaque to UV light. In addition, the area of exposure and the depth of penetration can be controlled by means of the EB conditions.

1.6.2.6. Conductive Adhesives

Conductive adhesives include both electrically and thermally conductive materials. Most adhesive fillers that result in electrical conductivity in a material also contribute to its thermal conductivity. However, thermally conductive adhesives that are electrically insulative are also available.

The majority of applications involve the use of silver, either in flake or powder form. This filler is preferred to gold because of its lower cost and its lower volume resistivity (that is, higher electrical conductivity). The best silver-filled epoxies have a volume resistivity of 0.001 Ω cm. With silver-filled adhesives, migration of silver to the surface under conditions of high humidity and direct current can occur. Neither gold- nor silver-coated copper fillers in various adhesives have this migration problem. Epoxies with up to 85 wt% silver filler are available, but adhesives with lower filler loadings generally have better strengths.

Other fillers that provide electrical conductivity include copper and aluminum, although oxide formation on the surfaces of these fillers can occur. This results in lower conductivity because of decreased particle-to-particle contact.

Recently, conductive epoxy resins and polyimides have been used as die-attach adhesives. The chloride ion contents of polyimides are inherently lower than those of epoxies. However, these adhesives are more expensive, have lower bond strength, and are more difficult to cure. Epoxies with low contents of ionic impurities, including chloride anions and sodium and potassium cations, have been available since 1981. These materials are fast curing and have excellent high-temperature strength and a glass transition temperature of up to 200 °C.

Oxide fillers are electrically nonconductive but are used to provide thermal conductivity. Alumina, the most commonly used filler, is fairly inexpensive. It can be added in high concentrations to epoxies and silicones without significantly increasing the viscosity of the uncured material. Minimum bond line thicknesses are desirable for thermally conductive adhesives because heat flow is proportional to the ratio of thermal conductivity to bond thickness. Alumina-filled epoxies contain up to 75 wt% filler and have thermal conductivities ranging from 1.38 to 1.73 $W\,m^{-1}\,K^{-1}$.

Other thermally conductive fillers include beryllium oxide, which is expensive and toxic, and boron nitride, which has a loading limitation of ca. 40 wt% in epoxies. Other inorganic oxides, including silica, are also used to provide thermal conductivity.

The chemical families that are used most often to provide electrical and/or thermal conductivity include epoxies, polyurethanes, silicones, and polyimides. Epoxies are the most widely used.

1.7. Bonding Techniques

Cleaning and Pretreatment of the Surfaces to be Joined. To insure good adhesion of the adhesives, the surfaces to be joined must be cleaned before the adhesive is applied. Layers of grease, oil, and lubricant are removed by means of solvents or aqueous degreasing baths. Loose dirt and oxide layers can be removed by brushing. In many cases, the surfaces to be joined are rubbed down with emery paper or sandblasted to remove firmly anchored layers that would interfere with the bonding process or to enlarge the surface by roughening. Pretreatment processes are used additionally to increase the affinity for bonding or (adsorption) activity of the surface and even to impart an affinity for bonding to certain plastics. In general, acidic pickling baths and often combined acidic and oxidizing pickling baths are used for metals, generating defined oxide layers. Alkaline pickling baths are used relatively infrequently. In plastics that are difficult to bond, the surface is modified by chemical pretreatment processes that increase the surface tension of the plastics and, hence, also their wettability and bondability [55]. The so-called corona treatment, i.e, electrical discharge in the presence of air, is used primarily for pretreating plastics films and, in some cases, also metal foils.

In addition to the cleaning and pretreatment of the surfaces to be joined, primers are applied before the actual adhesive in individual cases, their function being to improve the adhesion of the adhesives and/or favorably to influence the resistance to aging and corrosion of the bond.

Adhesive Application. Water-based adhesives and pigmented solvent-based types should be stirred thoroughly before application to remove possible inhomogeneity. For two-component or multipackage adhesives exact dosage and homogenous mixing of the components is essential. The method of application of the adhesive determines the uniformity of the bond line, the thickness of the adhesive layer, and therefore the

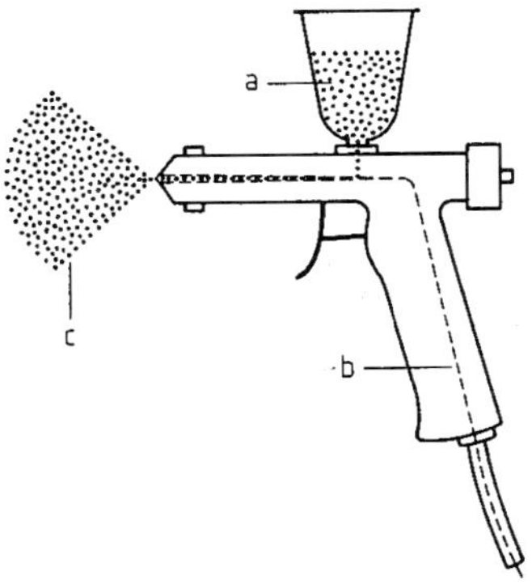

Figure 1. Adhesive application by means of a spray gun
a) Adhesive; b) Compressed air; c) Atomized adhesive

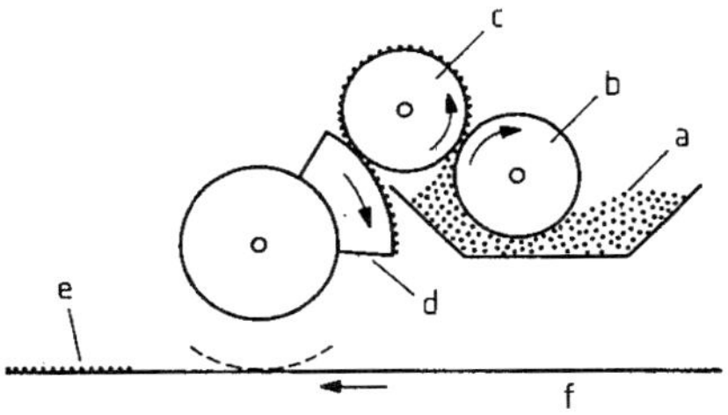

Figure 2. Application by segments
a) Adhesive; b) Dipping roll; c) Transfer roll; d) Rotating segments; e) Applied adhesive; f) Substrate web

quality and economy of the bonding process. The application method itself depends on the size and nature of the surface, the rheology of the adhesive, the required coating thickness, the number of components to be bonded and the speed of the process.

Manual application of the adhesives with toothed spatulas, blades, casting knives, or brushes is the simplest but least uniform method of application. Pasty adhesives can be applied from cartridges which are squeezed by hand or mechanical devices. This method is also used for moisture-curing polyurethane hot-melt adhesives with heated cartridges.

Low-viscosity adhesives are distributed over large areas by spray guns in the same way as paints (Fig. 1). The spraying process may involve the usual atomization with air or can be airless in the case of solvent-based adhesives.

Rollers are a versatile application method and range from simple glue rollers fed from a reservoir to highly sophisticated systems with several rolls, suitable for hot-melt, solvent-based, water-based, and solvent-free reactive adhesives (Figs. 2 – 6). The coating thickness is regulated by the distance between the rollers, the friction between them (different rotational speeds) and/or by blades. By this means, it is possible to apply from 0.5 g/m up to > 20 g/m in a very precise operation.

Gravure rollers are used together with doctor blades for a constant coating weight over the width in a narrow range of thickness. Unless the entire surface must be coated with adhesive, slim rollers or so-called wheel applicators are used. If the adhesive is only applied at intervals or with interruptions, segment rollers or wheels, interrupted rollers, toothed wheels, profile rollers, rollers with moving segments, or plungers are used. In

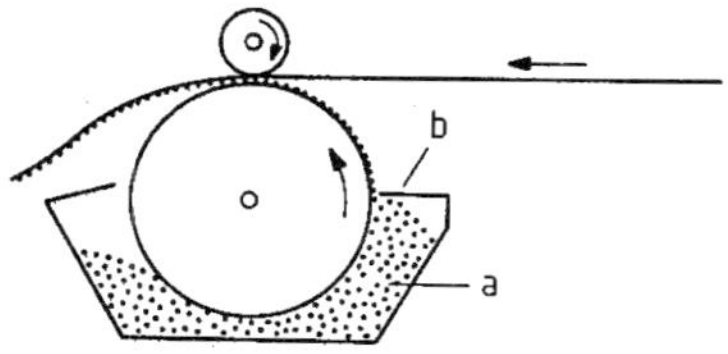

Figure 3. Application by rolls combined with a blade
a) Adhesive; b) Doctor blade

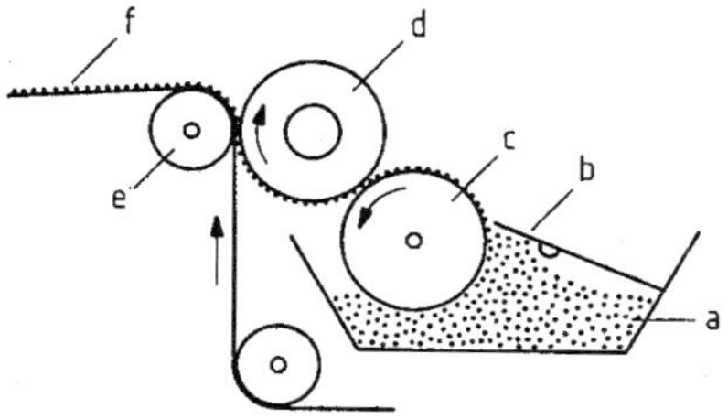

Figure 4. Application by a wheel applicator system
a) Adhesive; b) Blade; c) Dipping roll; d) Shaft with wheel applicator; e) Backing roll; f) Coated substrate web

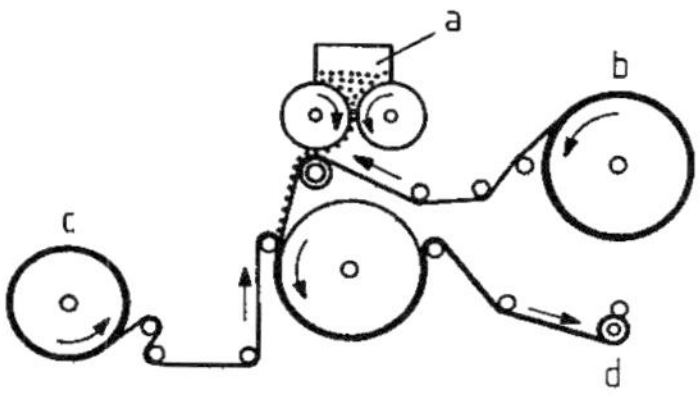

Figure 5. Coat application of hot-melts by rolls (System Pack 400/Kroenert)
a) Hot-melt adhesive; b) Roll I; c) Roll II (cover web);
d) Winding up roll

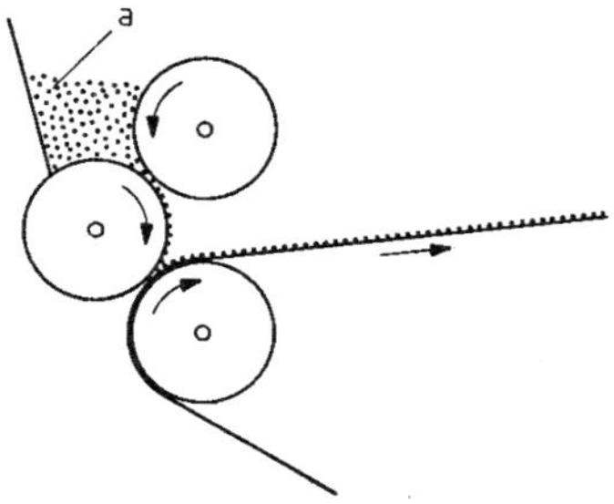

Figure 6. Reserve roll coating system (System Pagendarm)
a) Adhesive

addition, blade coating and airbrush coating are used. Hot-melt adhesives must be melted prior to application by rollers. The adhesive is usually transferred from the rolls to the substrates by direct coating in the machine direction. In some cases, reverse-roll coating (Fig. 6) has advantages, especially for low-viscosity adhesives.

Curtain coaters are used for applying adhesives over wide areas by passing the web below the nozzle of the tank (Fig. 7). Highly viscous hot-melt adhesives and thermoplastic polymers are applied from wide-slot nozzles to form a weblike film on the substrate.

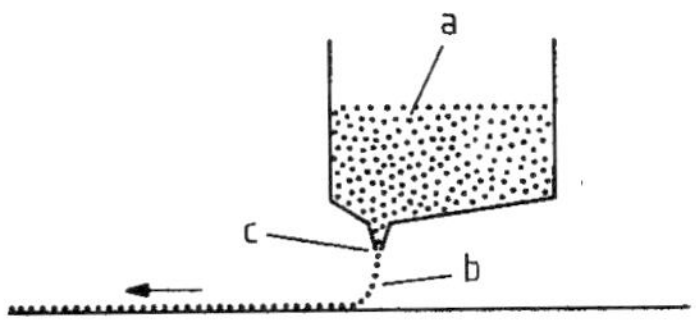

Figure 7. Curtain coater
a) Adhesive; b) Flowing adhesive "curtain"; c) Casting slot

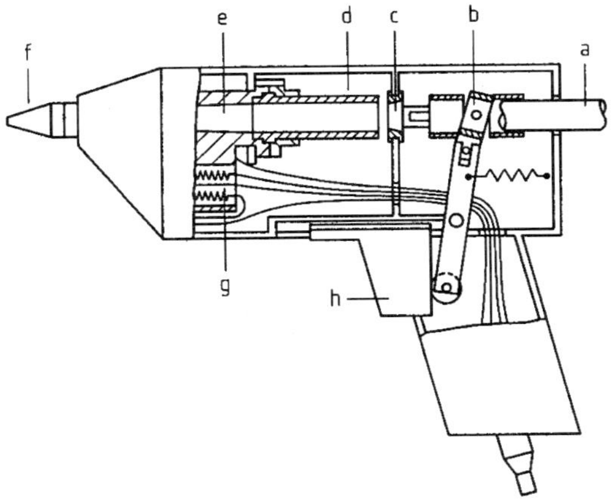

Figure 8. Hand gun for hot-melt (system lang)
a) Adhesive stick; b) Feeding; c) Retaining ring; d) Seal; e) Melting chamber; f) Nozzle; g) Wedge heater or heating wedge; h) Trigger

Typically, hot-melt adhesives for folder packaging are liquefied in melting tanks or drum melters and then pumped through heated hoses to the application nozzle. The piston pumps or gear pumps generate sufficiently high pressures to allow contactless application or spraying.

The porous coat system (Nordson) allows the application of discrete, random, and open patterns of hot-melt adhesive to substrates such as films, papers, fabrics, and nonwovens. With the control coat system, hot melt adhesives are applied continuously or intermittently by air-controlled nozzles without contact to the substrate. This technique is used as well for reactive hot-melt adhesives such as moisture-curing polyurethane hot-melt adhesives.

For temperature-sensitive substrates and prolonged open times, gases are incorporated in the melt under pressure. As a result, a hot-melt foam is formed during application [56].

The application units of solvent-free laminating machines consist of three or four rollers with differential speeds to lower the coating weight to 1.5 g/m^2, enabling machine speeds from 250 up to 500 m/min.

Relatively simple handguns (Fig. 8) are used for the workshop or do-it-yourself application of hot-melt adhesives.

Metering and mixing units (Fig. 9) of various constructions (gear pumps, piston pumps, static mixers) are used for the application of two-component adhesives and provide precise dosing and mixing even at elevated temperatures (70 °C).

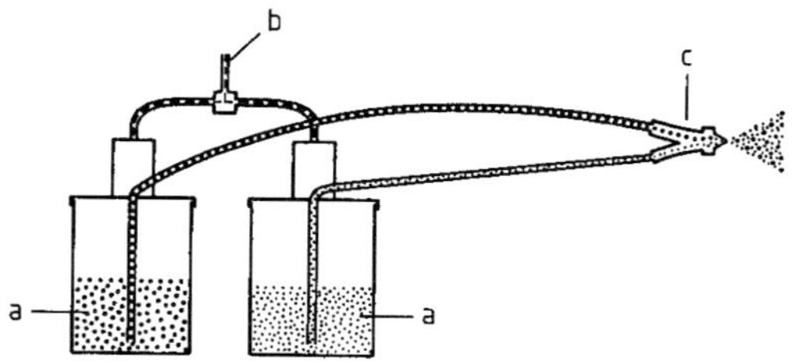

Figure 9. Two component mixing equipment a) Tanks with metering pumps; b) Compressed air; c) Mixing gun

Anaerobic reactive adhesives and cyanoacrylate adhesives are applied by means of special metering units that dispense small predetermined quantities at the required time intervals.

Joining, Fixing, Drying, and Hardening. After applying the adhesive to one or both surfaces, the substrates have to be fitted together within the open time and fixed until the bond is sufficiently firm. The pressure that has to be applied for bonding is determined only to a minor extent by the adhesive and largely depends on the geometry, nature, and deformability of the substrates and by its recovery forces.

The pressure for joining and fixing to give whole-surface contact can be very high, particularly in the case of large-area bonds, although the adhesive itself does not require such high pressures, apart from contact and hot-setting condensation adhesives. The necessary pressure is usually applied by pressure rollers, laminating rollers, or braces and presses.

The setting time and hence the fixing time can be shortened by heating. Warm-setting adhesives require certain minimum temperatures and times for curing. The energy required may be supplied by recirculating air ovens, drying tunnels in laminating machines and lacquering ovens, by radiant heaters, and by contact heating in heated presses or laminating rollers. Inductive heating of metal-to-metal bonds and high-frequency infrared heating of aqueous adhesives are also used occasionally. Within the last years UV-curing adhesives have been introduced, while electron-beam-curing adhesives are still in the development stage.

Polycondensation adhesives must be subjected during hot hardening to an external pressure that is higher than the vapor pressure of the cleavage products formed during condensation. Accordingly, these adhesives are applied in heated presses or by using pressure bags in heated ovens. The vacuum/pressure process in heated autoclaves is preferably used for uneven, three-dimensional surfaces [57].

A special case of bonding is the heat-sealing of previously applied heat-seal adhesives. Such coatings are converted by heat into a tacky state and bonded by the application of pressure. Thermal activation is carried out by heated sealing bars or plates, hot air, and sometimes by IR radiation.

High-frequency welding is used for special applications, in which cases the necessary heat is generated by the interaction of high-frequency radiation with the dipole moment of the pre-applied heat-seal adhesive itself.

1.8. Testing of Adhesives

The testing of adhesives comprises two aspects:

1) The adhesive as an individual compound
2) The performance in the actual adhesive joint that is formed between substrates

Generally the following adhesive properties are determined:

- Solids content
- Rheological properties
- Shelf life or storage temperature range
- Mechanical properties: tensile strength, modulus, Shore hardness, and elongation to break
- Glass transition temperature
- Softening point
- Flammability and flash point
- Health and environmental classification

More important, however, is the performance of the adhesive in an actual joint. Standard test specimens are bonded under standard conditions and subjected to destructive tests. The load values of these tests give information on the adhesive strength between two bonded substrates. Standard destructive test methods are:

- Tensile test (EN 26922)
- Flatwise tensile shear test (EN 1465)
- Peel test (EN 1464/ISO 4578 and EN ISO 28510-1,2)

Standard destructive tests are, above all, tensile, shear, and peel tests on specially made, strip-form test specimens (Fig. 10). In general, short-overlap test strips are used for tensile shear tests and flat-bonded test strips for peel tests. Bending and impact tests occasionally are carried out on these or similar test specimen geometries. Apart from the adhesive itself, the material of the test specimen, surface treatment of the specimen before bonding (sandblasting, pickling, priming), dimensions, overlap surface area, the test speed, the test temperature, and the direction in which the force is applied, all affect the outcome of these tests. Accordingly, the nature and dimensions of the test specimens, the overlap length, the peeling angle, and also the test conditions are laid down in test standards and other guidelines. The test standards [DIN, EN, ASTM, ISO, BS, PSTC (Pressure Sensitive Tape Council)] are designed to give comparable results. Shear strength and tensile strength are expressed in N/mm^2 (strain/joint area), peel strength in N/mm (strain/specimen width). In general, adhesive joints are more resistant to shear and tensile forces than to peel and bending forces.

A distinction can be drawn between:

- Static tests to failure under an increasing load
- Static long-term strain test below the breaking strain (creep)
- Dynamic tests with different loads and frequencies

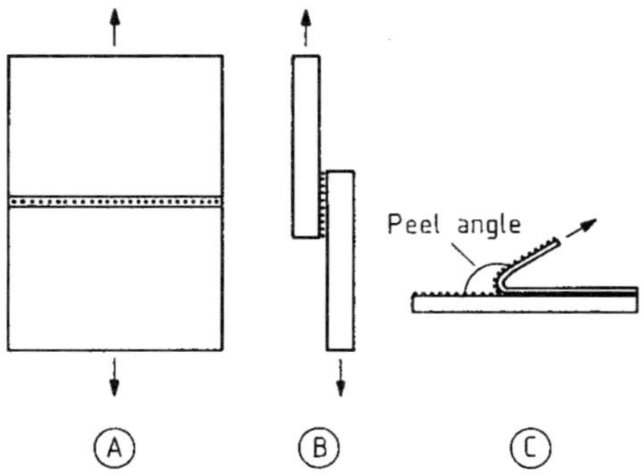

Figure 10. The most important kinds of strain for bonded joints
A) Tensile strain; B) Shear strain; C) Peel strain.

The static test can be carried out with standard tensile testers that cover the appropriate force range, optionally with a data recorder to obtain a force plot. Dynamic tests require special vibrating devices with constant loads. After the destructive test further information can be gathered from the test specimen. The nature of the break is an important indication of the quality of a bond line. There are three types of failure:

- Cohesive failure: break in the bond line
- Adhesion failure: break between substrate surface and adhesive (poor adhesion)
- Substrate failure, break in the substrate, the bond line is stronger than the substrate (structural bonding)

A combination of the above breakage patterns is possible. This is then indicated by a percentage figure (e.g., 60 % cohesive, 40 % adhesive failure). The above tests are also performed with exposure of the test specimens to extreme environments. Bonds are preferably tested for their resistance to:

- Natural weathering
- Climate variations (cycling)
- Heat aging
- Humidity aging
- Corrosive atmospheres
- Salt spray mist (automotive)
- Underwater storage
- Submersion in various chemicals such as petroleum, acids, and alkalis

The test conditions are generally much more severe than required for the actual final use of the bonded species. However these exaggerated test conditions deliver valuable information on the durability of the bonded structure in real life conditions. A good understanding has been developed for forecasting the bond line dependability on the basis of the results of aging tests.

The testing of specimens bonded under laboratory conditions is used to compare the performance of adhesives and substrates or the combination thereof, especially in quality-control laboratories. Under real-life conditions laboratory results may not be applicable. Surfaces of substrates, their geometry, curing temperatures, humidity, and many other factors may have an impact on the final bond quality. Therefore it is very important to perform a bond test in the final application even when the laboratory results were satisfactory. In many cases this is the only way to obtain information on the reliability of the bonded construction.

Testing of Anaerobic Adhesives. Testing standards and performance specifications for anaerobic adhesives and sealants have been established by government agencies and industrial organizations in several countries. In the United States there are military specifications for thread lockers, sealants, and retaining compounds. Many of these specifications are being superseded by commercial standards.

Mil-S-22473E, 12 April 1983 "Sealing, Locking and Retaining Compounds: (Single Component)" covers 15 of the earliest "Letter Grade" products. Specifications are set for color, viscosity, locking torque on 3/8-24 steel nuts and bolts and fluid tightness. The effects of immersion in a number of fluids, heat aging at 149 °C and hot strength at 149 °C (or 93 °C for some grades) are also measured. This specification calls for measurement of an "average locking torque" after 90, 180, 270 and 360° of turn.

Mil-S-46163A, 12 July 1983 "Sealing, Lubricating, and Wicking Compounds: Thread-Locking, Anaerobic, Single-Component" covers nine grades of product for sealing (Type I), lubricating (Type II), and wicking (Type III). Specifications are set for color, viscosity, locking torque (break and prevailing torque) on 3/8-16 steel, zinc- and cadmium-plated nuts and bolts, fluid tightness, lubricity, and "wicking" into preassembled fasteners. The immersion, heat aging, and hot strength tests are similar to those specified in Mil-S-22473E.

Mil-R-46082B, 10 June 1983 "Retaining Compounds Single Component, Anaerobic" (Amendment 6, 9 January 1990) covers three types of retaining compounds which are tested with a pin and collar compressive shear specimen. The three types vary primarily in viscosity although there are also some differences in heat resistance and strength. These products are subjected to immersion, heat aging and hot strength tests similar to those described above.

In Britain the Ministry of Defence specification DTD 5628-5633 covers test procedures and performance requirements for a range of products. Five strength bands and four viscosities from penetrating to thixotropic are defined. The torque strengths, including breakloose torque, are tested on M8 nuts and bolts and the shear strength in 12 mm pins and collars [201].

For an improved method of detecting the breakloose torque in a threaded component bonded with an anaerobic thread locker/sealer, see [202].

A British Standard BS 5292 has been prepared which relates to the use of anaerobic sealants on gas appliances.

In Germany standards have been published describing the "Compression Shear Test" (DIN 54452), "Dynamic Viscosity Determination of Anaerobic Adhesives by Rotational Viscometer" (DIN 54453), "Initial Breakaway Test at Bonded Threads" (DIN 54455) and "Torsion Shear Test" (DIN 54455).

In the United States the Industrial Fastener Institute has published standards for "Test Procedure for Locking Ability Performance of Non-metallic Locking Element Type Prevailing Torque Lock Screws" and "Test Procedure for the Locking Ability Performance of Chemical Coated Lock Screws." ASTM has published a "Standard Test Method for Shear Strength of Adhesives Using Pin-and-Collar Specimen" (ASTM D 4562-90, October 1990.)

ISO 10964 (August 15, 1993) "Adhesives — Anaerobic adhesives — Determination of torque strength of anaerobic adhesives on threaded fasteners" describes testing procedures for liquid and preapplied sealants.

1.9. Applications of Adhesives

It is possible here to select only typical examples for each subheading.

1.9.1. Bookbinding

Bookbinding with adhesives has played an important role since the beginning of the production of graphical products [185]. Originally, vegetable and animal raw materials were the main ingredients of adhesives, but a steady development of the adhesives took place to fulfill manufacturing requirements. Today, a range of adhesives, tailor-made for particular applications, is available. These applications include perfect binding, case making, casing in, back-lining and gluing-off of thread-stitched book blocks, as well as side-gluing and end-sheet tipping. The most important field is perfect binding.

Perfect Binding of Books. It is impossible to imagine the production of books, brochures and catalogues without perfect binding. New paper qualities such as high-weight coated (HWC) papers, lightweight coated (LWC) papers and recycling papers, high-quality papers such as matt-varnished papers and the increase of the production speed up to 300 books per minute require new bookbinding adhesives. Coated papers can have more than 50 % coat and therefore are difficult to bond. In general, the fiber structure of recycled papers is deteriorated by the recycling process, and these papers must be sized strongly, for instance, to withstand the pulling strain in offset printing, which causes a reduction of absorbency. Therefore, a longer setting time is needed for aqueous adhesives. Strong sizing can also lead to a change of the surface properties, which may cause adhesion problems. Occasional adhesion problems occur because of the often lavish designs of the graphic products. Varnishing, laminating, embossing, and other refinements have the result that the adhesives can not reach the paper fibers, which are otherwise easy to glue.

Four adhesive types are used for perfect binding:

- Glutin glues (animal glues)
- Emulsion adhesives

- Hot-melt adhesives
- Reactive polyurethane hot-melt adhesives

Glutin adhesives are used in form of jellies which are processed at 50 – 70 °C. Due to a sol – gel transition, very fast setting is achieved and a high production speed is reached. However, the adhesion spectrum of glutin glues is normally not very wide.

In contrast, many synthetic polymers possess excellent adhesion characteristics. For example, poly(vinyl acetate) homopolymers combined with external plasticizers show high adhesion, sufficient cohesion, and are easy to use [186]. The improved productivity of the machines has led to the development of fast-drying systems (by high-frequency, IR heating, or gas dryers). The book block can be further processed after cooling for 4 – 5 min.

Hot-melt adhesives are especially useful when high production speeds are required. On rapid cooling, they set fast and achieve the necessary strength quickly. Hot-melt adhesives based on ethylene – vinyl acetate (EVA) copolymers are most commonly used in the graphical industry, although occasionally other thermoplastic polymers are used. Reactive polyurethane (PUR) hot-melt adhesives have proved to be especially useful for high-quality perfect bindings [187]. PUR hot-melt adhesives, which have a very good adhesion characteristics and excellent strength, are applied at 120 – 130 °C. After cooling (physical setting), they react with moisture (chemical setting) to give high-molecular compounds which are practically unmeltable. Besides good adhesion characteristics, they have high chemical resistance (e.g., to the oil of printing inks), high temperature resistance, and good aging resistance. PUR hot-melt adhesives require special premelters and application equipment because of their reactivity. The PUR adhesives combine the advantages of dispersions (e.g., good adhesion, high aging resistance) and of hot-melt adhesives (e.g. high production speeds) in a single system.

Another method of optimizing adhesion and cohesion is to use two different adhesives, one immediately after the other (two-shot application) [188]. The two adhesives can be of the same type (e.g., two dispersion or two hot-melt adhesives), or two different systems can be used (e.g., a dispersion primer followed by a hot-melt or PUR hot-melt adhesive).

Case Making. In case making, when the covering material consists of textiles, plastics film, or paper, as in the backlining of books, hot animal glues are used because they set rapidly by gelatinization and rapidly bond the narrow overlap of the coating material around the board, even under the brief contact pressure normally applied in the machine. In special cases, also emulsion-based adhesives are also used.

Casing In. Depending on the case material and the machine speed, normally emulsion adhesives [poly(vinyl acetate) or copolymer-based] with good flowing properties are used for casing in.

1.9.2. Adhesives for Packaging and Packaging Materials

Manufacture of Corrugated Board. For this purpose starch suspensions are used almost exclusively. Some of the starch is gelatinized under the heated corrugating rollers which form the corrugation, thus insuring high initial tack. The use of emulsion-based or hot-melt adhesives is confined to special cases. Hitherto, it has been discouraged by the relatively high price of these products, which only occasionally is compensated by other advantages.

Lamination of Paper and Board. Adhesives based on starch, dextrin, glutin and poly(vinyl alcohol), and also emulsion adhesives, mainly poly(vinyl acetate), are used for the lamination of paper and board. Only when high initial tack and a lay-flat effect are required are hot glues based on glutin still in use. Starch-based adhesives modified with silicate, casein-stabilized copolymer dispersions, and rubber latices are used in the lamination of paper to aluminum foils. Modified polyacrylate solutions and reactive polyurethane adhesives are used mainly for paper-to-plastic lamination.

Adhesive Coating of Paper and Board. Pressure-sensitive coatings, for example, for self-adhesive labels and envelopes include natural rubber – resin combinations in solvents, polyacrylate emulsions, and hot-melt adhesives. Polyacrylate emulsions and hot-melt adhesives are becoming increasingly significant because they are solvent-free.

Heat-sealable coatings are applied from emulsions, solutions, or melts and are activated in different ways: heat-sealable coatings with delayed tack mainly by heat in the absence of heavy pressure, and typical heat-sealing coatings by pressure and heat applied to the bond area as contact or high-frequency heat.

Moisture-reactivated coatings are produced from dextrin glues, dextrin emulsions, mixed glues, and remoistenable hot-melt adhesives.

Bags, Sacks, Pouches, and Boxes. Starch-based adhesives and also emulsions of poly(vinyl acetate) homopolymers and copolymers or mixtures thereof are used in the manufacture of bags, pouches, and carrier bags, depending upon the type of materials (e.g., paper, plastic films, etc.), surface treatment, and the speed of the machine. Machine design factors necessitate the use of pastelike starch-based or emulsion-based adhesives for the bottom gluing of block-bottom or cross-bottom bags.

In making paper sacks, starch-based adhesives normally are used for the longitudinal seam and also for the bottom and bottom patching. For paper sacks with PE free film or inliner, emulsion-based synthetic resin adhesives are used. For the longitudinal seam of the inliner, hot-melt adhesives are used.

The manufacture of folding boxes has long been a domain for free-flowing, low- to medium-viscosity emulsion-based synthetic resin adhesives.

Envelopes. In general, the top-flap gumming of envelopes is made with modified conventional dextrin glues and dextrin – emulsion mixed glues, which dry more quickly and show better ability to stay flat for automatic enveloping. To an increasing extent, modified emulsions and remoistenable hot-melt adhesives are being used. Dextrin glues and synthetic resin emulsions normally are used for bottom-flap and window gluing. Gums for self-seal gumming are based on natural rubber. Self-adhesive top flaps are based on polyacrylate emulsions.

Adhesives for Automatic Packaging Machines. In automatic packaging, package production, filling, and closing are often carried out in a single machine. For this reason, very fast setting adhesives are used.

Considerable recovery forces occur in the closing of prefabricated boxes, particularly in seaming machines, so that hot-melt adhesives often have to be used in such cases. Emulsion-based adhesives are normally used for the dust-tight sealing of cases. Slide boxes, deep-freeze packs, etc., can be line- or spot-sealed with hot-melt adhesives.

Cartons are sealed with emulsion-based or hot-melt adhesives, depending on the length of the pressure belt and the quality of the cardboard. On modern, fast-running machines with up to 120 cases to be closed per minute, hot-melt adhesives are preferably used, due to their fast setting. An interesting combination is the simultaneous use of emulsion-based and hot-melt adhesives. In this case, a spot application of hot-melt adhesive performs the function of the pressure unit or compression section, whereas the actual bonding function is performed by emulsion-based adhesives, generally applied in line form.

Cigarette Manufacture. For gluing cigarette seams, depending on the application system, highly viscous or free-flowing starch glues or emulsion-based adhesives are used. Hot-melt and emulsion-based adhesives are used for cigarette-filter seams, since their viscosity, initial tack, and setting properties are compatible with the special conditions of this manufacturing step, both at low startup speeds and at maximum speeds of 600 m/min. Emulsion-based adhesives are used for bonding the filters to the cigarettes.

Adhesives for Laminated Films. Laminated films are currently produced by two techniques: lamination and extrusion. In lamination, the low-viscosity adhesive solution is applied by rolls to one of the films in a thickness of a few micrometers and dried. The two films are then combined in the laminator. The adhesives used are either two-component or one-component, moisture-cross-linking polyurethane adhesives, depending on requirements (type of film, sterilization resistance, sealing seam strength, etc.). Besides the traditional solvent-based systems developments have been made in the direction of solvent-free systems (high-solids products) and of polyurethane emulsion-based systems.

In extrusion, a polymer melt is extruded onto a primary web. An improvement in adhesion or reduction in the extrusion temperature is obtained partly by applying an adhesive as primer to the primary web. Adhesives suitable for this purpose are similar to those used in a lamination. In coextrusion, two polymer melts are combined with one another. If required, a polymer primer is extruded between the two melts.

Bonding of Cellophane. Cellulose hydrate film is used for bag manufacturing or for wrapping operations. Because it readily takes up moisture in its unlacquered state, aqueous emulsion-based adhesives are used primarily for bonding. The addition of glycerol ensures they release their water only slowly so that the film stays flat. Lacquered cellophane is either bonded or sealed. The adhesives do not require any additives to keep the film flat because the film is protected by the lacquer. The adhesives are normally based on poly(vinyl acetate) copolymers or other flexible polymers. Where solvent-based or hot-melt adhesives are used, both types of film may be bonded with the same adhesive, because the absence of water eliminates the problem of keeping the film flat.

Manufacture of Polyethylene Sacks. The one-component and two-component polyurethane adhesives used for the production of laminated films (see above) are used for this purpose, depending on the type of machine.

Labeling of Glass. For labeling glass, dextrin, starch, and casein glues are used. With many machines, the adhesive used is required to show pickup ability, that is, the adhesive applied, for example, to a pallet must be tacky enough that it can pick up a label from the label magazine. Furthermore, the label must not slip after application, and any tendency to curl as a result of moistening must be counteracted. Dextrin glues are well endowed with these properties. In addition, by virtue of their high solubility in water, any glue that remains on the machine is easy to clean off. However, dextrin glues cannot be used where resistance to condensation, water, or ice is required (e.g., in breweries). Therefore, casein glues are mainly used in breweries. Although they are not as concentrated as dextrin glues (ca. 60 %), they develop a high initial tack, even on wet bottles, because they undergo a marked increase in viscosity on contact with the cold bottle. An additional reason for the use of casein adhesives is their wider adhesion range compared to dextrin adhesives. The latter adhere reliably to normal glass bottles, but their adhesion to coated or surface-treated glass bottles is not sufficient, so that casein adhesives have to be used.

Modern automatic high-performance labeling machines are capable of labeling between 500 and 1300 bottles per minute. At lower speeds, starch glues, which do not show such pronounced pick-up properties, are still used because of their lower cost. However, their higher water resistance can cause problems in bottle-washing machines for returnable bottles.

Labeling of Plastic Containers. The glues normally used for glass bottles cannot be used for plastic containers because they do not adhere sufficiently to plastic. Plastic bottles normally consist of polyethylene, polypropylene, poly(ethylene terephthalate), polycarbonate, poly(vinyl chloride), or polystyrene. Canlike containers and tablet tubes are made of polystyrene or polypropylene. Modified copolymer emulsion-based and hot-melt adhesives are used for labeling.

Although there are no major adhesion problems with PVC, PET, and PS, adhesives with stronger adhesive properties must be used for polyethylene and especially polypropylene. However, if the polyethylene surface is treated (e.g., exposed briefly to a gas flame) before labeling, it is possible to use the same adhesives as for PET or PVC.

1.9.3. Nonwovens Hygiene Industry

In general two types of adhesives are used for the manufacture of baby diapers, feminine-hygiene, and incontinence products. For simple construction purposes, hot-melt adhesives based on amorphous poly(alpha-olefins) (APAOs) are used. High-quality constructions are produced with adhesives based on styrenic block copolymers. For bonding of elastic materials such as polyurethane foam, lycra and natural rubber strands, and hydrophilic and hydrophobic of nonwovens, pressure-sensitive adhesives (PSAs) based on styrenic block copolymers are used. Hot-melt adhesives are the most common types of adhesive. The most widely used synthetic block copolymers are styrene – butadiene – styrene (SBS), styrene – isoprene – styrene (SIS), and styrene – ethylene – butadiene – styrene (SEBS) with different styrene contents.

The adhesives are mainly applied intermittently (up to 650 pieces per minute for baby diapers and up to 1000 pieces per minute for feminine-hygiene products) at line speeds of up to 400 m/min. These requirements place high demands on the performance properties of the adhesives.

Construction Adhesives. APAO-based adhesives are mainly applied by slot nozzles on both the polyethylene backsheet and the nonwovens top sheet. Usual application temperatures are 125 – 140 °C.

The thermoplastic rubber-based adhesives are applied by application methods such as spiral spraying, fiberization, and melt blowing. Application temperatures are usually in the range of 140 – 165 °C.

Elastic Attachment Adhesives. These thermoplastic rubber-based adhesives are usually applied by spraying at 140 – 165 °C. As different elastic materials are used, different adhesives need to be formulated to give the right adhesion performance for each substrate.

Positioning adhesives for feminine-hygiene products are usually thermoplastic rubber-based, pressure-sensitive, and have low viscosities. The tack level can vary from very low to very high depending on the type of product and customer requirements. The positioning adhesive can be applied by both slot and fiberization applicators. The application temperatures depend on the type of substrate and vary from 120 to 170 °C.

1.9.4. Hygienic Papers

Four processing steps for the production of hygienic papers, (e.g. toilet paper, kitchen rolls) involve the use of adhesives:

Tube Winding. Mainly dextrin adhesives are used and/or plasticized homopolymeric poly(vinyl acetate) emulsions, but glutin glues are also used in special cases.

Pick-up adhesives bond the tissue to the Kraft paper tube. Depending on the machine systems the following adhesives are used: If the adhesive is applied directly to the tube by a slice (machine speed up to 400 m/min), aqueous colloidal solution systems at room temperature are normally used. If the adhesive is applied directly to the tube by a nozzle (machine speed up to 450 m/min), aqueous polyol-based adhesives are often used at room temperature.

When the adhesive is applied by a slice to the application roller, which transfers the glue to the tube (machine speed up to 650 m/min), hot-melt adhesives are used at 60 – 80 °C.

For high speed machines (up to 850 m/min), in which the adhesive is applied directly as a strip by means of a nozzle (temperature of the nozzle 80 °C), aqueous colloidal solution systems are used.

Laminating Adhesives. Aqueous adhesives based on cellulose or poly(vinyl alcohol) for the ply lamination of tissue papers must give clean machinability, good initial tack, and reliable bonding strength of the tissue laminates.

Tail Tie. For this application aqueous adhesives based on cellulose or poly(vinyl alcohol) are used. For the tail tie of tissue paper rolls, products with high wet tack, transparent film, and a low final bond strength are necessary.

1.9.5. Gluing of Wood and Wooden Materials

Today for the assembly gluing of wood, for gluing veneers, plastic sheets, and films, and for the manufacture of wood-based materials (chipboard, plywood, hardboard, profiles), synthetic adhesives are used almost exclusively. The traditional use of adhesives based on natural products (glutin and casein glues) is confined to a few special cases only (for example, violin making).

Poly(vinyl acetate) emulsion "white glues" are used widely for the gluing of wood and wood-based materials. They generally contain small quantities of solvents and/or plasticizers to adjust the film-forming temperature and chalk as a pH buffer and filler. They have a solids content of approximately 50 – 65 % and a film-forming temperature of 0 to 15 °C. Special components may be added to increase the setting rate or to prolong the open time. The application weight is 100 – 200 g/m^2, the open time 5 – 25 min, and the clamping time 5 – 45 min for a compressive force of 5 – 20 N/cm^2. The exact figures strongly depend on the glue coating weight and the type of substrate.

Bonds with higher water resistance can be achieved by using special blends or two-component emulsion glues. The second component may be a diisocyanate or particularly an inorganic salt (usually, aluminum nitrate). These hardeners, added in a quantity of at least 5 %, improve the moisture resistance of the joint. However, their addition limits the pot life.

Normal glues develop EN 204/205 Class D2 strengths which are adequate for all indoor applications, for example, for the assembly gluing of furniture and for laminating plastic sheets. Self-cross-linking one- or two-component poly(vinyl acetate) glues, which develop Class D3 and D4 strengths according to the same European standard, may be used for gluing in moist atmospheres (bathrooms) and for exterior assembly gluing (windows, doors).

Furniture Foil Adhesives. These adhesives consist primarily of ethylene – vinyl acetate copolymer emulsions. Rigid PVC and melamine – paper foils, generally 0.2 – 0.4 mm thick, are applied to wood-based materials for decorative purposes. The adhesives must exhibit high tack and high resistance to heat and cold and must counteract the shrinkage of the film (application weight 60 – 100 g/m^2).

Hot-melt adhesives are mainly used for automatic edge-banding applications to bond edge materials to particle board, as well as for profile wrapping with foils, paper, and veneer.

Increasingly, hot-melt adhesives also are being used for the postforming process. Immediately after surface lamination, the overhanging laminate is bonded to the contoured panel edge without an additional adhesive joint.

Hot-melt adhesives are applied on edge-banding machines, profile-wrapping machines, and postforming machines. The adhesive is applied to the substrates with special rollers or slot nozzles. The joint between the cover and the core material is achieved by applying pressure with a roller system while the adhesive sets.

In assembly processes, such as furniture or door manufacture, the hot-melt adhesive is applied in spots or lines by hand or with automatic guns. Often the hot-melt adhesive is exclusively used as an additional assembly support due to its high initial strength, while poly(vinyl acetate) glue is used for achieving the final strength (two-shot process). Hot-melt adhesives for the edge-banding and postforming process are highly viscous products, often filler-free or with a filler content of up to 50 %. For profile wrapping, low-viscosity hot-melt adhesives are used, sometimes with fillers. For assembly bonding, a wide range of assembly hot-melt adhesives varying in viscosity, setting time, and open time is available for different equipment and uses.

As adhesives based on ethylene – vinyl acetate copolymers cover a very wide range of applications, they are used in large quantities. In the case of higher demands on the bond, polyamide- or polyolefin-based hot-melt adhesives are recommended. Highest bonding quality regarding heat and cold resistance, bonding strength, and waterproofness are achieved by using polyurethane-based, moisture-reactive hot-melt adhesives.

EVA hot-melt adhesives have heat resistance up to 85 °C, polyolefin and polyamide hot-melt adhesives up to 130 °C, and reactive polyurethane hot-melt adhesives up to 150 °C. The cold resistance lies between – 20 and – 30 °C; in the case of polyurethane-based hot-melt adhesives, it is even lower than – 30 °C.

Urea formaldehyde (UF) resin glues are the most widely used adhesives in woodworking and are preferably applied by the hot-glue technique. They are marketed as precondensates in the form of 50 – 66 % solutions or as dry powders that must be dissolved in water. Curing aids can be added. In veneering, contact times as short as 30 s are possible for press temperatures of 100 – 120 °C. The most common additive is ammonium chloride. The amount of hardener governs both setting time and pot life. Glue joints in thick boards can be set by high-frequency heating.

For gluing plywood, the UF resins are generally mixed with extenders such as rye flour, followed by the addition of curing aids, the type and amount of which governs the setting characteristics at the processing temperature and pressure. The glues thus prepared are applied to the dry veneers (moisture content 6 – 12 %) by a glue spreader (application weight 120 – 200 g/m^2), and the glue-coated veneers are assembled and then glued in multilayer hot presses or high-speed single-layer presses at 100 – 160 °C and 60 – 200 N/cm^2, depending on the type of wood. For particle board production, unextended UF resins are used preferably. The particles are coated with glue by spraying, the particles of the center layer being coated with less resin than the particles of the two surface layers. Average glue resin content is ca. 10 % (i.e., 6 – 8 % for the center layer and 8 – 12 % for the surface layers). The glue-coated particles are first spread to form a three-layer cake, precompressed, and then compressed in hot presses to form particle board.

UF glues ideally satisfy the requirements for interior use but are unsuitable for exterior application. The requirement for low formaldehyde content must be considered by all producers of particle board. As plywood and veneer glues, with regard to water resistance, UF resins satisfy the requirements of DIN 68705 and EN 314, category IF, and as particle-board binders they meet the requirements of DIN 68763, category V 20.

The glues are cheap, easy to apply, and therefore suitable for the mass production of wood-based materials.

Melamine resin glues are important in the woodworking field when weatherproof glue joints according to DIN 68705, category AW at relatively low press temperatures (under 100 °C) are required. With added extenders moisture-proof glue joints can be obtained that meet the requirements of the same DIN category IF, at press temperatures above 75 °C.

Melamine–Formaldehyde (MF) Resins. The largest field of application for melamine formaldehyde (MF) resins is the manufacture of decorative laminates and the surface finishing of chipboard, hardboard, and profiles. Special papers with weights of 18 – 150 g/m^2 are impregnated with MF resins, dried, and then pressed onto the substrates at 130 – 180 °C. Their colorless transparency, hardness, chemical resistance, and absence of yellowing have made these resins particularly suitable for surface finishing. For technical applications including, e.g., concrete form-work boards, they are also applied by the wet-resin process, in which the resin solution is applied to the surface and cured under heat and pressure.

Phenolic resin glues, for cost reasons generally derived from phenol cresol mixtures, are used for the hot gluing of plywood, the manufacture of chipboard, and as binders for fiberboard. They are marketed in liquid form. Their shelf life is limited to about six months. They set at 130 – 180 °C, generally with addition of hardeners. For setting at room temperature, the addition of a special hardener (usually *p*-toluenesulfonic acid) is essential. The use of phenolic resins requires a relatively low moisture content of the wood (6 – 10 %).

Joints based on phenolic resin glues are of high quality and weatherproof. They are recommended for gluing wood to metal (metal-clad wood).

Resorcinol resin glues set without the assistance of heat under mildly alkaline conditions. Therefore, no damage is caused to the wood by acid hardeners, and there are no restrictions on the moisture content of the wood up to 18 %. The resins can be cured with paraformaldehyde. Wood-to-wood glue joints with resorcinol resins are among the strongest and are weatherproof and resistant to solvents, oils, and fats. For reasons of economy, mixed condensates of resorcinol and phenol are used instead of pure resorcinol.

Contact Adhesives. In countries with a high standard of mechanization, contact adhesives based on polychloroprene rubber for woodwork are used only for on-site repairs and for special purposes, e.g., for gluing plastic sheets or veneers onto rounded substrates. Here, contact adhesives are widely replaced by poly(vinyl acetate)-based wood glues. However, in less developed countries, contact adhesives are still widely used for veneering and laminating plastic sheets to wooden substrates. Their advantage over all other wood glues is the high bonding strength achieved virtually immediately after joining under short-term pressure (several seconds) without the need for sophisticated equipment.

Polyurethane Adhesives. The search for formaldehyde-free binders resulted in a new generation of polyurethane-based glues for chipboard production. The usage of polyurethane based adhesives for assembly of wooden materials has also increased. These liquid polyurethane adhesives are single-component adhesives that harden by a chemical cross-linking reaction which is initiated by moisture in the material. The water resistance of wooden joint fulfills the requirements of EN 204, D4. Furthermore, the adhesive joint shows high resistance to a variety of organic solvents, weak acids, and alkalis.

Over-all Quality Standards for Wood Glues. There are several national and European standards for the adhesives used in woodworking, of which the following are the most important: EN 204, EN 205, DIN 53255, DIN 68705, and DIN 68761 – 765.

1.9.6. Footwear Adhesives

One of the most important methods of assembling shoe components is adhesive bonding. In 1949 polychloroprene adhesives replaced the nitrocellulose solutions that had been introduced in 1906. The almost complete versatility of the polychloroprenes allowed leather, textiles, and other natural substances (until then the only shoe materials) to be supplemented by other materials, such as various rubbers and synthetics, or to be replaced altogether, as in the case of toe puffs, counter stiffeners, or heels. The way was thus opened for an astounding development in fashion and shoemaking techniques. Adhesive-based processes, supported by new machines and equipment, led to new bottom-attachment systems, simplified the production of footwear, and drastically reduced the number of production operations. Since 1970 a position of comparable importance has been reached by polyurethane adhesives.

Other adhesives for shoe manufacturing are solvent adhesives based on natural rubber or SBS or SIS block copolymers, dispersions of polyacrylates and poly(vinyl acetate), and a range of latex and hot-melt adhesives.

Closing Room Adhesives. These adhesives are light-colored and nonstaining solutions (or latices) of natural or synthetic rubber; other polymer dispersions are also possible. Wet-stick and pressure-sensitive bondings predominate. Also PSA hot-melt adhesives are of increasing importance for these applications. If higher strengths are demanded of the bonds, for instance, for replacing stitched structural seams, the requirements are met by two-way contact adhesives or by adhesives weldable in the high-frequency process. Ethylene – vinyl acetate hot-melt adhesives are used for applying backings and for sticking-in toe puffs and counter stiffeners, whereas polyamide hot-melt adhesives are used for folding, edge binding, including french binding, etc.

The components to be bonded (linings, interlinings and other reinforcements, and also toe puffs and counter stiffeners) can be precoated with adhesive and then ironed in.

The Lasting Operation. Apart from the bottom attachment, the lasting operation produces the most vital structural bond in the shoe. It involves the fastening of the upper material over the margin of the insole and is subjected to considerable stress;

there is normally no supplementary stitching, tack-driving or stapling. A common lasting system is combined pulling-over and cement-lasting with automatic injection of hot-melt adhesive; this is mainly a rapid-setting crystalline polyester or the more flexible polyamide, both supplied in rod or granule form. Lasting by contact bonding with polychloroprene adhesives (or in light shoes also with natural rubber solutions and latices) is limited to particular constructions.

Bonding Soles to the Lasted Uppers. Polychloroprene and polyurethane solvent adhesives, occasionally also dispersions of the same polymers, are used for this purpose. Various pretreatments are used to improve the bond: application of a thin first coat of adhesive, especially for absorbent surfaces; priming, in particular the halogenation of vulcanized rubber and thermoplastic rubber soling; the use of base-coat primers for nylon and other synthetics, as well as UV-curing primers for EVA and other substrates in athletic footwear.

Depending on the required productivity, quality, and performance a wide variety of processes are used in sole bonding. As an example for a contact bond, the single-component adhesive is applied to the sole, usually in advance. The soles can be stored (dust-free) until required in the production process. On the lasted margin of the shoe upper, the adhesive is usually a two-component mix containing isocyanate hardener; bonding must be carried out within the specified open-assembly time.

For bonding with heat activation, the sole is activated by IR irradiation or rapid flash activation: the activation time may be 2 – 6 s, and a surface temperature 55 – 80 °C can be reached. During the process it is the adhesive film that is heated, rather than the sole, which remains cooler. This ensures a more rapid development of bond strength. Soft soles that tend to deform when hot can be bonded at room temperature or with adhesives that have a low activation temperature.

For the various molding-on processes, such as the direct injection molding of thermoplastic rubber or PVC, the vulcanizing-on of rubber, and the low-pressure reaction molding of integral-skin cellular polyurethane, adhesives can seldom be dispensed with: the lasted margin of the upper is coated with a polyurethane adhesive, often containing a pre-mixed latent hardener, or with a self-vulcanizing synthetic rubber solution compounded to match the vulcanization system of the soling compound.

With moisture curing polyurethane hot-melt adhesives a new innovative technology has been introduced in the sole-bonding process. It can be used as a one-way bonding system or as a two-way contact adhesive, applied either to the sole or to the upper. In this process, an adhesive coat must be applied on the other substrate.

Adhesives for Shoe Repairs. These adhesives must be almost universally applicable, because widely differing shoe types and material combinations must be repaired. For this reason, one-component polychloroprene-based adhesives are used almost exclusively, though for PVC and some other plastics materials polyurethane adhesives are quite common.

Testing. In Europe adhesive properties are tested according to European standards (CEN) EN 522, EN 1391, and EN 1392. Specified standard test materials can be obtained from the various research institutes of the European shoe industry.

1.9.7. Bonding of Plastics

Plastics are bonded to one another or to other materials. They differ widely in their affinity for bonding. Some plastics are easy to bond (e.g., PVC-U, ABS, PS), whereas others can be bonded only after special preparation (e.g., PTFE). On account of the low polarity and surface tension of certain plastics (e.g., polyolefins) there are still no adhesives for pipe joints which are durably resistant to shear strain and pressure. The choice of adhesive is determined by the substrate and the mechanical, thermal, and chemical requirements [69] – [75].

Poly(Vinyl Chloride). Rigid PVC (PVC-U) and chlorinated PVC (PVC-C) pipes, sheets, and films are used worldwide in the construction industry. They are joined by adhesives consisting of PVC solutions in a mixture of solvents, which operate by the principle of diffusion bonding. The bond is established by diffusion of the adhesive into the surfaces and the temporary dissolution of the PVC (cold swell welding).

For the bonding of uncalibrated pipe joints, adhesives with gap-filling properties are used, in which PVC of a certain molecular mass is dissolved in a solvent consisting predominantly of THF, cyclohexanone, or *N*-methylpyrrolidone. The viscosity range is 2000 – 3500 mPa s. The application time of the adhesive ranges from 1 to 5 min. The adhesives are preferably made thixotropic so that they do not run off the surfaces to be joined after application and to avoid the damage to the plastic material because of accumulation of adhesive (the excess solvent can cause tensile corrosion). After preliminary cleaning of pipe and fitting with solvent (cleaning and dissolution) and application of the adhesive to both sides, the pipe and the fitting immediately are fitted together without turning. The bond is established by diffusion of the adhesive into the surfaces and temporary dissolution of the PVC (diffusion bonding).

PVC adhesives, usually consisting of solutions of post-chlorinated PVC in methylene chloride, are used in the case of inorganic acids as flow-through media in PVC pipe because of their better chemical resistance. Since PVC adhesives have no gap-filling properties a special application technique is required. This priming method is little used today.

Contact adhesives based on polychlorobutadiene or polyurethane are used for large-area bonds between PVC-U and PVC-C sheets and wood or metal.

Vessels are often lined with thin PVC-U and PVC-C films (1 – 2 mm thick). Contact adhesives are used with usual application techniques. For bonding cooling tower films, special adhesive solutions, which are harmonized in viscosity and thixotropy with the processing techniques (roll or pot application), analogous to PVC pipe adhesives, are used.

Adhesives for bonding plasticized (flexible) PVC (PVC-P) films are adhesive solutions based on THF (80 – 90 %) and PVC-P (10 – 20 %). Adhesives based on nitrile rubber or polyurethane in a plasticizer-resistant formulation for the bonding of PVC-P films made from differently formulated compounds are also suitable.

Adhesives based on polymethacrylates and vinyl copolymers are used for the continuous bonding of PVC-P films to metal sheets (coil coating). In this process, the adhesive is applied by machine to the cleaned sheets, the solvent is removed, and the film subsequently is rolled into the bed of adhesive heated to 140 – 180 °C.

Polystyrene, Styrene – Acrylonitrile Copolymers, and Styrene – Butadiene Copolymers. These plastics may be bonded to themselves or to one another simply by treating the surfaces to be joined with pure solvents, such as dichloromethane, ethyl acetate, or 2-butanone. However, this method of bonding, which also is known as solvent welding, does involve the risk of material corrosion. Therefore, it is advisable to use solutions of polymer raw materials in mixtures of organic solvents with a low risk of material corrosion. Depending on the field of application, especially in combining different materials, contact and nitrile-rubber adhesives containing aliphatic hydrocarbons as solvents also may be used. Pipes of ABS/ASA (ASA: acrylonitrile – styrene – acrylic ester polymer) for high-temperature wastewater run-off systems (max. 90 °C) preferably are joined with solutions of ABS/ASA in acetone or 2-butanone, although they also may be bonded with THF – PVC systems. The solutions are of medium viscosity (max. 2000 mPa s) and have a predominantly thixotropic rheology.

Solutions of ABS in 2-butanone are preferred for use in pressure applications, such as compressed-air conveying systems. As with PVC, the bond is established by diffusion.

Rigid polystyrene foam is used for the manufacture of composite sheets, for example, for containers and prefabricated building boards. The rigid foam is combined with metals, wood-based materials, asbestos, cement, and plastic sheets. Polyurethane adhesives are particularly effective in this field.

Polyolefins. Polyethylene, polypropylene, and polybutene can be bonded only after treatment to increase the surface energy, generally by oxidation, and make the surfaces receptive to bonding. Pretreatment can be carried out with an oxidizing flame, with oxidizing chemicals, or by electrical discharge.

Nowadays, the flaming process is rarely used: in the lining of containers with polyethylene and in the printing of bottles (Kreidel process). Among the chemical pretreatments, immersion in chromosulfuric acid (for a few seconds to a few minutes), followed by careful rinsing with water has proved successful. Thin plastic layers for the production of laminated films for use as packaging materials are best pretreated by corona discharge.

The pretreated polyolefins can be bonded with contact adhesives, epoxy resin adhesives, one- and two-component polyurethane adhesives, and hot-melt adhesives.

Polyamides are bonded with concentrated (ca. 85 %) formic acid. Two-component polyurethane or epoxy resin adhesives are suitable for bonding to metals, for example, for holding bearing bushings in place.

Polymethacrylates can be bonded by treating the surfaces to be joined with a mixture of dichloromethane and dichloroethylene. However, this does involve the risk of material corrosion. In such cases, polymerization adhesives cured by light can be used. Polyurethane, epoxy, and contact adhesives are also suitable.

Fiberglass-Reinforced Plastics. To obtain high-quality pipe connections, the surfaces to be joined must be roughened and the pipes carefully tapered. Polyester adhesives or thermosetting epoxy resin adhesives are best suited for bonding. A pressure of 50 – 100 N/cm^2 is applied to the joined surfaces.

Fluoropolymers. There are no known adhesives for fluoropolymers that give bonds of adequate strength without pretreatment. Simple, low-quality bonds can be obtained with contact adhesives. After chemical pretreatment, for example, with a solution of an alkali metal naphthalenide in THF, polyfluorocarbons can be bonded with high strength by using epoxy and polyurethane adhesives.

Curable Molding Compounds. High-strength bonds involving these plastics are obtained with epoxy resin adhesives. The adherents must be roughened lightly and thoroughly cleaned beforehand. Laminates, including decorative laminates, of phenolic resin- or melamine resin-impregnated papers or fabrics can be bonded with contact adhesives.

1.9.8. Bonding of Elastomers

In the rubber industry, a fundamental distinction is drawn between "bonding" and "adhesion". Whereas the adhesion of generally vulcanized rubber articles to one another or to metals or plastics is solely a function of the adhesive used, bonding is a process in which vulcanization of the rubber and, at the same time, adhesion between rubber and substrate are obtained by the action of heat and pressure.

1.9.8.1. Adhesion of Rubber

Contact Adhesives. Nowadays, rubber solutions based on natural rubber or styrene – butadiene rubber, optionally in conjunction with regenerate and depolymerized rubber, are used for large-area bonding where adhesion does not have to meet stringent requirements, as is the case, for example, with the lining of vessels, containers, etc. with rubber.

Today, multipurpose adhesives based on polychloroprene rubber generally are used as adhesives for rubber articles.

Reactive Adhesives. Epoxy and polyurethane adhesives are suitable for bonding polar elastomers, such as nitrile rubber and chloroprene rubber. However, epoxy adhesives form a relatively rigid bond so that they are suitable only for bonding to rigid substrates. In contrast, polyurethane adhesives form elastic bonds and therefore are used for rubber-to-rubber and rubber-to-leather bonding, for example, in the manufacture of profiles and in the shoe industry.

Polyurethane adhesives also are suitable for bonding nonpolar elastomers, for example, natural rubber, styrene – butadiene rubber, or ethylene – propylene terpolymers, after chemical pretreatment of the surface.

Polymerization adhesives based on α-cyanoacrylates (see p. 30) are suitable for special bonds involving rubber, because they enable bonds of high strength to be obtained quickly and easily. Medium-viscosity types are preferred for rubber-to-rubber bonds by virtue of the minimal hardening of the joint. In the rubber industry, for example, the adhesives are used for bonding profiles to one another and also for repair work.

Because of the increasingly stringent requirements that rubber articles have had to satisfy over the past 20 years, the bonding process is preferred in the manufacture of rubber – metal composites.

1.9.8.2. Bonding of Rubber

Rubber-to-Rubber Bonding. In the bonding of unvulcanized rubber mixtures to one another, there is generally no need to use adhesives when the mixtures to be combined with one another are identical or similar in composition. However, adhesives have to be used for combining different types with one another. Various bonding agents based on halogenated polymers are available commercially for this purpose and also may be used for bonding vulcanized to unvulcanized articles.

Rubber-to-Fabric Bonding. Rubber-to-fabric bonding is of considerable commercial significance in the tire industry and industrial rubber goods such as conveyor belts and rubberized fabrics. In the tire industry, the so-called resorcinol – formaldehyde (RF) latices are still used predominantly for bonding rayon, whereas mainly resorcinol – formaldehyde – vinylpyridine latices are used for bonding nylon and adhesively finished polyester yarns, which, unless specially pretreated, adhere very poorly to rubber.

In industrial rubber goods, which exhibit considerably greater variations in rubber composition, solutions of unvulcanized rubber mixtures are used. They are applied to the fabric as a first coat with the addition of polyisocyanates to improve adhesion. Directly adhering rubber mixtures containing resorcinol – formaldehyde donors and an acidic silica filler are also used for the production of fabric-reinforced rubber materials.

Rubber-to-Metal Bonding. In the Ebonit or hard rubber process [76], the bonding agent is a reaction product of natural or synthetic rubber with 30 – 40 % sulfur, which is thermoplastic in character, adheres firmly to metal, and provides for good adhesion to soft rubber. This process is still in use today in the lining of vessels and, to a certain extent, in the manufacture of large-diameter rolls.

The brass process makes use of the fact that electrodeposited brass of certain composition and crystal structure is capable of directly bonding rubber mixtures, providing their composition is compatible with the adhesion process [77]. Whereas today the brass process is of only minor significance for the production of rubber – metal composites, it is widely used in tire manufacture for bonding the bead wires and for bonding the carcass mixture of steel cord.

Certain polyisocyanate adhesives allow bonding of rubber to metal with far less technical effort than by the brass process. However, polyisocyanates are extremely sensitive to atmospheric moisture and react with compounds containing OH and NH_2 groups, such as antiagers. This reaction leads to a weakening of adhesion.

Nowadays, rubber – metal composites are produced with bonding agents in which the polymer base is usually a halogenated polymer, for example, a mixture of chlorinated and brominated polybutadiene. After the film of bonding agents applied to the metal component has dried, the metal component is bonded to the rubber mixture during the vulcanization process.

Mechanical and/or chemical pretreatment of the metal is essential for corrosion-resistant bonding (for the effect on bonding of the blasting medium used for blasting, see [78]).

The fillers in the rubber have no significant effect on bonding. However, soft rubber mixtures (30 – 40 Shore A) are more difficult to bond than hard rubber mixtures because of their high plasticizer content. The degree of vulcanization is another determining factor as far as the quality of bonding is concerned [79]. Bonding errors generally occur if the mixture is scorched before it comes into contact with the bonding agent surface or if the mixture is undervulcanized or severely overheated [80].

1.9.9. Bonding of Metals

The adhesives used for bonding metals are preferably reactive adhesives, predominantly epoxies, phenolics, (meth)acrylates, polyurethanes, although poly(vinyl chloride) plastisols, MS polymers, and rubber adhesives also are used for elastic bonds [81] – [85]. The range of bond strengths obtainable extends from high-strength, structural bonds (tensile shear strength ca. 40 N/mm^2) to highly elastic adhesive sealing compounds (tensile shear strength ca. 1 N/mm^2). However, the strength of a bond depends to a large extent upon the nature and direction of the forces acting upon it and on the temperature. Because the intrinsic strength of the cured adhesive resins is not comparable with that of metals, provision must be made for sufficiently large joint areas. Accordingly, the design of adhesive joints in metals must be appropriate for bonding. Resistance to aging and corrosion depends on the nature of the surface pretreatment and on the type of adhesive used.

Epoxies, polyurethanes, and (meth)acrylates represent the principal types of adhesive for structural bonding of plastics and metals. These technologies compete for various applications in the transportation and marine industries because of their strength, good environmental durability and cost effectiveness. Historically, urethanes were chosen for low-temperature applications because their hard segment/soft segment structure typically allows them to retain flexibility in cold environments. Conversely, epoxies were preferred for high-temperature environments because of their strength and high cross-link density. Methacrylates fill the gap between these two technologies. They offer good resistance to high and low temperatures, and they provide the distinct advantages of rapid curing at room temperature, less sensitivity to mixing ratio and higher tolerance of surface contamination. They are proven performers in the transportation and marine industries. Typical applications include bonding fiber-reinforced plastics to metal.

The aeroplane industry was the first to incorporate metal bonding in assembly procedures on a significant scale. The increasing use of high-strength lightweight metals which cannot be satisfactorily welded and the reduction in overall weight through the introduction of light-gauge constructions were other important influencing factors. Light-gauge sheets (skins for fuselage and wings, control surfaces, etc.) could be stiffened more effectively by cementing onto corresponding profiles than by riveting. The problem of reinforced openings (windows, door frames) and of distributing stresses over large cross-sectional areas (rotor blades) was satisfactorily solved by multilayer construction, especially supporting-core and sandwich constructions, the widespread application of which actually was made possible by bonding. In this method, high-shear cores of various structures (e.g., honeycomb), are bonded to thin cover layers or skins of aluminum and other materials. This gives a lightweight, nondenting, torsion-resistant construction suitable for aircraft floors and wings. Suitable adhesives are phenolic–poly(vinyl formal) and epoxy adhesives and also epoxy–nylon adhesive films.

Vehicle manufacture (automobiles, railroad vehicles) is another significant application for metal-to-metal bonding.

In *lightweight metal constructions,* tubes and hollow profiles of any cross section can be designed and bonded as socket joints. Door and window frames are made from steel and aluminum profiles with angles bonded in place (epoxy adhesives), a heat-insulating intermediate layer being applied by bonding or made by casting with an adhesive (polyurethane, epoxy adhesives).

In *bridge building,* adhesive joints (epoxies, polyesters) are used in combination with high-strength, on-torque threaded fasteners (to absorb peel forces) for load-bearing steel constructions.

In the *electrical industry,* it is above all the bonding of sheet packs (dynamo sheets, transformers, motors) and the fixing of small parts (ferrite cores) that are increasing in significance (epoxy resins, cyanoacrylates). Adhesives (epoxy resins) containing conductive additives also are used, for example, for printed circuits.

In *machine construction* and *instrument* and *tool manufacture,* plain bearings, axle bearing guides, bushings, etc., are made by bonding, and punches are cemented into stamping tools with epoxy resins, cyanoacrylates, and acrylate adhesives. Screws, gear wheels, shafts, etc., are secured with anaerobic adhesives.

Adhesive pastes are used as *repair kits,* often in combination with woven glass-filament cloth for reinforcement. In this way, bodywork is repaired, worn surfaces are renewed, and pipes are sealed (epoxies, unsaturated polyester resins).

The sandwich construction method mentioned above involves joining *metals to other materials* by bonding. Further examples include the bonding of brake linings (phenolic adhesives) and compound materials in ski manufacture, where aluminum is bonded to plastics, wood, etc. (phenolic and epoxy adhesives). Highly alloyed steels, beryllium and titanium alloys, and other special metals can be bonded with adhesives (e.g., polyimides, polybenzimidazoles) that have comparable high-temperature resistance.

1.9.10. Adhesives for Wallcoverings

Wallcoverings, classified by EN 233 – 235 [133], consist of a variety of materials: paper, textiles, glass, plastics (PVC, PE, PS, PU), ceramics, wood, cork, metals, leather. They are attached by hand in the form of sheets (wallpapers), tiles, or profiles to masonry, plaster, concrete, plasterboard and gypsum wallboard, wood, and metals.

For ease in do-it-yourself projects (adjustability, open time, easy application, etc.) and because most substrates to which the coverings are bonded are capable of absorbing moisture, it is preferable to use water-based adhesives in the form of solutions and emulsions. Solvent-containing adhesives are only used on nonporous substrates, for example, metals [86], [87].

The adhesives are based on methyl cellulose, starch ethers, carboxymethyl cellulose, pre-gelatinized starch, poly(vinyl acetate), polyacrylate, styrene – butadiene latices, polychlorobutadiene, and polyurethanes. The viscosity of the adhesives covers the range from fluid to pasty.

Wallpapers include those with a PVC or metal surface and rough-textured plain wallpapers. They are applied with pastes of pre-gelatinized starch, starch ethers, carboxymethyl or methyl cellulose in the form of 2 – 10 % solutions. For rough-textured plain wallpapers and heavy wallpapers, moisture resistance and tack are increased by utilizing poly(vinyl acetate) redispersion powders. High solid, emulsion-based adhesives, optionally in conjunction with resin-enhanced pastes (special purpose pastes), are used preferably for heavy vinyl- and metal-faced wallpapers.

Light *textile fabric wallcoverings* are hung with special fabric wallcovering adhesives. Special-purpose pastes containing 10 – 20 % of emulsion-based adhesives also are used.

Heavy wallcoverings of unbacked PVC (1000 g/m^2 or more) can be applied with adhesives based on acrylate or poly(vinyl acetate) copolymer emulsions. The same applies to *metal foils* that are applied underneath wallpaper as a vapor barrier.

Polystyrene and *polyurethane foam* and *mineral wool tiles* are applied to interior and exterior for decoration and insulation purposes (sound and heat). Solventless adhesives based on poly(vinyl acetate), polyacrylate, or styrene – butadiene copolymer emulsions are preferred for this purpose. The viscosity of the adhesives must be adjusted to give the required wet tack.

Ceramic wall tiles are applied either with cement mortar adhesives or with adhesives based on styrene – butadiene copolymer or polyacrylate emulsions. Tile adhesives must be pasty and thixotropic to prevent the tiles from slipping but allow positioning.

Wood panels, chipboard, and similar wooden materials can be fixed with aqueous adhesives based on poly(vinyl acetate) or polyacrylate emulsion. The adhesives must be pasty and have high initial strength. In some cases they are applied from cartridges. Also contact adhesives, special reactive systems (hybrid silicone), and hot-melt adhesives are utilized.

1.9.11. Floorcovering Adhesives

Flooring adhesives are used for attaching floor coverings to a variety of substrates. The adhesives must develop a strong and durable bond between the floor covering and the substrate. They should not adversely affect coverings, supports and substrates, and after application, should remain odorless and have low emissions. Usually flooring adhesives are applied to the substrate by hand with trowels or sometimes rollers.

Recent environmental and occupational-health legislation has resulted in a major shift from organic solutions to aqueous emulsions of polymeric compounds [88], and the use of waterborne adhesives is accepted as the state-of-the-art for most floor covering installations. Waterborne flooring adhesives are often based on acrylic copolymers, vinyl acetate – acrylics, vinyl acetate – ethylene – acrylics, or styrene – butadiene rubber latex. They commonly contain natural (e.g., colophony and colophony esters) or synthetic resins (e.g., hydrocarbon resins) and inorganic fillers. In recent years the content of solvent has been reduced to below 5 wt %. The use of high-boiling solvents (glycol ethers, liquid hydrocarbons) has been adopted to some extent for the formulation of solvent-free flooring adhesives [89]. There is now a move to eliminate high-boiling substances and even traces of volatile contaminants from the adhesives [90]. Several European manufacturers have established a chamber test method to record VOC and to classify low-emission products for flooring [92], [203]. Today solvent-based products are only needed for special applications (e.g., water-sensitive substrates).

Linoleum Adhesives. Linoleum generally is laid with waterborne adhesives based on acrylic emulsions or SBR latex. They contain higher amounts of inert inorganic fillers, and natural and/or synthetic resin is used to accelerate initial tack. Linoleum adhesives have to set rapidly to avoid expansion or shrinkage of the covering. Two-component adhesives consisting of one polymer dispersion and a second cement preparation are available to adhere linoleum to nonabsorbent substrates.

Parquet adhesives can be cold- or hot-spread compositions. The former can be classified as waterborne, solvent-borne, or reactive. The first two types are commonly based on poly(vinyl acetate) emulsions in water or as solutions in organic solvents. Reactive systems are solvent-free two-pack polyurethane adhesives or epoxy resin adhesives. They are preferred for all parquet and wood types which are sensitive to water. All adhesives are required to show high cohesive strength but certain plasticity to absorb movements of the wood through shrinking and swelling in response to changes in atmospheric humidity without failure of the joint. Hot-spread adhesives, which are bituminous or tar-based, are increasingly being replaced by the cold-spread types.

Adhesives for Plastic Floor Covering. Solvent-free adhesives based on acrylic and vinyl acetate – ethylene – acrylic emulsions are used for bonding vinyl, synthetic rubber, and polyolefin flooring. Solvent-containing contact adhesives (rubber- or polychloroprene-based) still are used for some applications (e.g., stairs). When water resistance is required, reactive polyurethane or epoxy resin adhesives are preferred to waterborne emulsions.

In addition to developing a certain minimum strength, the adhesives primarily have to meet demands on the dimensional stability of the covering.

Because of plasticizer migration between covering and adhesive joint, the adhesives for plastic floor coverings should be adjusted carefully in their composition. Plasticizers are commonly used for the production of coverings based on PVC.

Adhesives for Textile Floor Covering. The composition of waterborne adhesives for bonding wall-to-wall carpets is similar to that of linoleum adhesives. High initial tack (wet tack) and a resistance against carpet cleaners is required. Previously used solutions of poly(vinyl ethers) and rosin in mixtures of highly volatile solvents are still used for carpets that are extremely sensitive to water.

Conductive Adhesives. Conductive adhesives are available for bonding conductive floor coverings. The electrical properties are achieved by the addition of carbon fibers, carbon black, or graphite. Conductivity of a floor covering is sometimes required to avoid undesirable static charging.

Tile Adhesives. The thin-set process uses hydraulically hardening thin-set mortars based on cements, emulsion adhesives based on aqueous synthetic resin emulsions, which can be single-component ready to use tile adhesives or two-component adhesives consisting of modified aqueous emulsions and a cement-based compound, and reactive adhesives based on epoxy resins or polyurethanes.

Depending on the intended field of use, cement-based tile adhesives are modified with different amounts of redispersible synthetic resin powder such as co- or terpolymers based on vinyl acetate or acrylates. The recipe must be adjusted to the different types of tiles and natural stones. On account of the various types of tile adhesives corresponding standards may be found, for example in DIN, BSI, AFNOR (Association

Française de Normalisation), UEAtc (Union Europeenne pour l' Agrement Technique dans la Construction) or as European Standards prepared by CEN.

1.9.12. Building Construction Adhesives

The use of adhesives in building construction imposed exacting requirements upon quality and application techniques [92] – [98]. In addition to durability and resistance to mechanical stress, temperature and weather resistance are prime requirements. In prefabricated construction, the use of adhesives was an important requirement for efficient assembly techniques.

Adhesives as Additives for Hydraulic Compositions. Synthetic resin dispersions and redispersible synthetic resin powders that are added to cement-containing concrete mixes and mortars must be resistant to alkali and dilutable with water. Redispersible powders are added to cement-based adhesives in factory made mortars, while synthetic resin dispersions are mixed with the powder compound on the building site just before the adhesive is applied. The products that have been specifically developed for this purpose include, in particular, copolymers of vinyl acetate in combination with other esters or with ethylene and terpolymers of vinyl acetate, vinyl chloride, and vinyl laurate and also copolymers based on acrylates, methyl methacrylates and styrene, as well as styrene – butadiene. The additives modify the properties of the concrete. For example, they considerably improve the adherence of new concrete to old concrete. In addition, resistance to dynamic stress is considerably increased. The addition of anionic melamine resins is recommended also for establishing certain rheological properties. Adhesives based on cellulose ethers are added to cement-based systems and gypsum plasters to improve adhesion to the substrate and, by virtue of their hydrophilic character, retain water, thereby preventing mortars and plasters from drying out and improving their strength. Starch and starch ethers are used to modify the adhesion of cement mortars to mineral substrates.

Prefabricated Construction. Sandwich elements are widely used in prefabricated construction. In this case, bonding must meet stringent requirements with regard to stability and durability, especially in the case of exterior wall elements. Deflections attributable to differential temperature or moisture-induced expansion or contraction can amount to as much as 5 % of the length. The adhesive layer must be permeable to water vapor or act as a water barrier, depending on the structure and function of a given element. The adhesives used are two-component reactive resins, based mainly on epoxy resins, but polyurethane resins are used as well, as are elastomer-based contact adhesives that are applied on both sides.

With building components based on synthetic resin concrete (polymer concrete), the cement may be replaced partly or completely by organic binders, such as co- or terpolymeric vinyl acetates, acrylates or epoxy resins. Such systems are described as polymer cement (PC) if the cement as binder has been totally replaced with organic binders, and as polymer cement concrete (PCC) if the cement binder has been partly replaced by organic binders. These systems are mainly used to reface concrete surfaces.

Low-viscosity reactive resins are used to seal cracks in concrete constructions. These epoxy and polyurethane resins show excellent creepability and bond strength to concrete.

Assembly Adhesives. In carcass work, cement mortar is still widely used as a binder for mineral components. The disadvantages of cement mortar include its long curing time, its poor adhesion to old concrete, and its poor resistance to dynamic stress. For these reasons, adhesive mortars have been developed. They are mixtures of liquid, curable synthetic resins, such as epoxy resins or unsaturated polyester resins, blended with mineral additives, generally quartz, and the corresponding hardeners. They establish an excellent bond with hardened concrete, stone, metals, and certain plastics, and are used for applications involving particularly severe stress, for example, for cementing rail supports onto concrete sleepers, for fixing dowels and anchor bolts in concrete, for steel reinforcements, and for the permanent bonding of high-grade concrete elements.

For applications such as swimming pools and water-purification tanks, for which chemical resistance is required, or where resistance to dynamic stress is needed, special epoxy resins are used. Blends of a waterborne synthetic emulsion with a cement compound are suitable for applications with lower requirements. In some cases, e.g. sewage-treatment plants, elastomer-modified bitumen mortars are still used.

Another important application of reactive resins is the production of resin-bonded screeds that, by virtue of their high mechanical and chemical stability, are used, for example, for industrial floors. One particular problem is the bonding of old concrete to new concrete, where peeling occurs because of the shrinkage of the new concrete layer. By applying intermediate layers of a suitable epoxy resin mortar, it is possible to establish permanent bonds with strengths greater than that of the concrete.

1.9.13. Adhesives for Bonding Textile Fabrics

Fabric bonding is a relatively new application for adhesives [99]–[103]. The particular demands that a fabric bond must satisfy make the replacement of conventional thread stitching very difficult:

- Adequate bond strength
- Resistance to washing and cleaning
- No staining or discoloration
- No spoiling of texture
- Maintain breathability (in the case of spread adhesives)
- Rapid setting

Adhesives are used widely in the apparel industry for applying interlinings. The interlinings or nonwovens are precoated with hot-melt adhesives. To maintain breathability, the adhesive is applied spot-wise rather than spread. Suitable adhesives for these and similar purposes are, for example, copolyamides of the 6,6/6,12 type and also boiling-resistant low-pressure polyethylene. 6,36/6,12/6-Terpolyamides based on dimeric acids have been used recently because of their high flexibility, resistance to washing, and favorable application properties. Their softening range is generally from 100 to 150 °C. After cutting to size, the precoated interlining materials are fused using ironing presses (front fixing).

The introduction of other inexpensive thermoplastics into the fabric bonding field has met hitherto with little success. Polyesters, for example, are not sufficiently resistant to dry cleaning. By contrast, polyurethanes show high resistance and favorable bonding properties.

Another area in which adhesives are very widely used in the textile industry is the manufacture of chemically bonded nonwovens. The nonwovens are preformed by the dry-layer or wet-layer process and subsequently bonded by spraying or impregnation with adhesives. The binders used generally are products based on polyacrylates, synthetic rubber, and vinyl polymers. Nitrile and styrene – butadiene rubbers are used for industrial nonwovens because of the need for resistance to oils, fats, and organic solvents.

1.9.14. Flocking Adhesives

Flocking adhesives are used for bonding short (0.3 – 5 mm) monofilament textile fibers (known as flock or flock material) to a variety of substrates [104]. Compared to mechanical flocking processes, in which the flock is scattered or blown onto a layer of adhesive, electrostatic flocking currently has the greatest importance. In electrostatic flocking, suitable finished fibers are oriented in an high-voltage electrical field and accelerated to such an extent that they penetrate deeply at one end into the open adhesive coating and are firmly anchored therein. The surplus, nonanchored flock is blown or brushed away after drying and setting of the adhesive.

Whereas flocking was formerly carried out for decorative reasons (imitation suede or velvet), technical and economic aspects (noise insulation, friction reduction) also have become significant. Major improvements in flocking adhesives undoubtedly have contributed to this situation. In addition to paperboard, woven fabrics, yarns, films, and wallpapers, plastics, plastic foams, metals, rubber profiles, glass bottles, and, occasionally, even interior walls and car bodies are being flocked to an increasing extent. The adhesives used for flocking primarily are emulsion-based adhesives, generally based on poly(vinyl acetate) or polyacrylates and their copolymers, in some cases with added cross-linking agents. However, solution adhesives (solutions of synthetic resins) and also solventless and solvent-containing reactive adhesives (epoxy and polyurethane adhesives) are also used.

The adhesives must satisfy various selection criteria, of which adhesion, strength, elasticity, and stability of the set adhesive film are the most important. In many cases, colored adhesives also are required. The adhesives generally are applied by spray guns or rolls. They must have sufficiently long open times and spread evenly on the substrate.

Because the adhesive represents the positive pole (earth) in the electrostatic flocking process, through which the charge must be dissipated rapidly, it is required to show a certain conductivity (maximum resistance 10^7 Ω) This is not a problem where water-based adhesives are used, but in the case of solvent-containing adhesives, the necessary conductivity often can be obtained only by means of additives.

1.9.15. Adhesives for Bonding Glass

The bonding of mineral substrates generally requires high-polarity, low-shrinkage adhesives. These two requirements are satisfied by reactive adhesives [105] – [111]. In many cases, the bonding of glass additionally requires high transparency and UV stability of the adhesive for optical reasons. The adhesives primarily used are polar polymers such as epoxy resins and polymethacrylates in the form of two-component systems. For large-area compound bonding on an industrial scale, for example, for automobile windshields and bullet-proof glass, glass sheets are combined with laminates of poly(vinyl butyral) and plasticizer. The permeability of transparent glasses to radiation also allows the use of UV-hardening one-component systems based on methacrylates. Elastic glass-to-glass bonds and similar bonds between glass and other materials are obtained with rubber adhesives based on silicone and polysulfide.

The epoxy resins occupy a leading position as glass adhesives. Their hardening at room temperature can be varied over a wide range. Different curing systems are available for this purpose. Thermosetting epoxy resins that harden in a few minutes at 120 – 150 °C are used on industrial scale.

In certain cases, improvements in adhesion can be obtained by additionally priming the glass surface. The primers used are bifunctional products, for example, organosilanes. The adhesives used for bonding utility glass also are required to be resistant to boiling water. Organosilicon epoxy resins have proved to be effective for this purpose, and are used either on their own or in conjunction with conventional epoxy resin adhesives. In addition to their high resistance to water, siloxane – epoxy resin adhesives are of relatively high thermal stability.

1.9.16. Adhesives in Automobile Manufacture

Introduction. A variety of different adhesives, sealants, and coating compounds with a wide spectrum of properties, from high-strength structural adhesives with energy-absorbing properties (crash resistance) to highly elastic surface-coating compounds, are used in modern automobile manufacture. The boundary between adhesives and sealants is fluid and cannot be defined precisely. It is intended here to discuss only products that are applied for bondings and/or sealings between two parts or in gaps (e.g., in weld seams) which are able to transmit constructive strength, not for surface coating purposes, such as underbody protection.

Although the quantity in which the adhesives thus defined are used per vehicle depends largely on the model, the average value in Europe appears at present to be between about 8 and 12 kg.

For adhesives used in the body shop, strength and elasticity are tested from −40 °C up to +90 °C, whereas aging behavior is investigated under temperature load, in humidity, or in aerosols of salt solutions (salt spray test). In recent years, there has been a steady increase in the use of alternating climate tests either with regular cycles between dry cold and humid heat (VDA) or with a single change from prolonged storage in humid heat to dry cold.

Because automobiles are manufactured in large numbers of up to a few thousand units per day, the adhesives must be formulated in regard to their processability such that they fit into the established production flow and its cycle time. In view of the increasing advance of automation and the use of robots for body assembly, ease of pumping through pipes, often of considerable length, and also appropriate rheology at the application stage are essential requirements. As far as possible, it is preferred to use one-component formulations with wash off resistance that are cured by the heat of the electrocoat, primer, and paint bake ovens. At the trim line, where no bake ovens are available, mainly one-component, moisture-curing or contact adhesives are used. Two-component products will acquire higher significance for bonding of plastics and for assuring rapid through cure. For new production concepts, the cure speed of trim-shop adhesives will be very important.

Body and Paint Line. On the body line, all adhesives and sealants are applied to the steel surface covered with oil that is used for deep drawing and for corrosion protection. Accordingly, the adhesives and sealants used must be able to absorb the oil. In addition the adhesive must bond to a large number of surface coatings such as zinc, Zn−Ni, electrolytic zinc, fire zinc, prephosphated Zn, and organic coatings.

As hem-flange adhesive on hood and bonnet and on doors, high-strength adhesives are used in order to provide these components with the necessary inherent stability. High impact performance imparts improved crash resistance to the bonded parts and safety for the passenger cell. In addition, the adhesives afford effective protection against corrosion in these particularly sensitive areas (PVC plastisols, acrylics, epoxy resins). For further improving protection against corrosion and to obtain a cleaner appearance, the closed hem-flange seam is sealed with a sealing bead of PVC or acrylic plastisols; some automobile manufacturers are changing their sealing concept (clean paint shop), and these operations are allowed only in the body and/or trim shop. In the hem-flange seam, adhesives have largely replaced the hitherto customary spot welds.

For joining anti-flutter panels to the outer panels of bonnet and trunk, adhesives of relatively low strength and high elasticity are used both at the assembly stage and after electrocoating (PVC plastisols, rubber-based adhesives). These properties are necessary to insure that the bond seams do not "show up" on the outer surface (read-through effect). In this area, adhesive bonding has paved the way for constructions that cannot be obtained by other joining techniques. Rubber-based tapes or compounds are used for joining roof bows to roof panels.

For better protection against corrosion in spot-welded seams, the seams are treated additionally with an elastic adhesive/sealant (spot-weld sealer) based on rubber. Hot- and cold-application systems are in use.

Sealing tapes and/or sealing compounds (rubber-based, vulcanizing with different degrees of expansion or permanently plastic) are used for sealing in the region of fenders, wheel housings, sliding roofs, air louvers, tank supports, and roof linings. Die cut pieces (rubber-based) are used in some areas (e.g., door hinges, mirror mounts). Highly expandable (500 – 1000 %) rubber or EVA-based materials are used for sound absorption in cavities.

To give washing out resistance, pasty adhesives must be solidified by pregelling or precuring before they enter the cleaning process by passing them through a recirculating air oven, IR heaters, or induction heaters. Cleaning is followed by pretreatment (phosphating), electrocoat application, and oven baking at 180 °C. Various weld seams (drip rails, inner seams, outer seams) are later sealed with PVC or acrylic plastisols. The next step is the application of surface primer and final paint with the associated bake ovens.

Trim Line. In the trim shop, the fresh paint surface forms the substrate for a number of bonding operations in which parts prefabricated to a greater or less extent and subsupplied parts are adhesively bonded into the painted body. Because no more bake ovens are involved, cold-curing systems (moisture or two-component systems) are employed.

At the glazing stage (windshield, rear, and side windows), the insertion of a tacky sealing cord (rubber-based) into the elastomeric profile surrounding the glass panel provides for better sealing. In the last 20 years, there has been an steady increase in the use of direct glazing techniques in which a moisture-curing, one-pack polyurethane adhesive or MS-polymer-joins the glass panel to the painted body. Advantages include very effective sealing, the possibility of larger screens, and lighter body constructions (the glass panel now contributes to body rigidity via the modulus of the adhesive), and more creative freedom (lower air resistance). Most recently adhesives for radio antennas integrated into the rear screen have been developed with high electrical resistance. Fast setting materials eliminate any fixing aids and guarantee short drive-away times.

Contact adhesives (polychloroprene, nitrile, or polyurethane rubber) are used for fitting vinyl roofs, roof linings, sound-deadening mats (where they are not already incorporated in the primer-coated body), and carpets. More recent prefabricated roof constructions are designed for bonding with special emulsions.

The waterproofing polyethylene film behind the door panel is sealed with pasty compounds or tapes (rubber-based). Folded lining films and hems on doors and carpets are sometimes secured with hot-melt adhesives (polyamide, polyethylene) because they set relatively quickly.

The mechanical spot fastening of the buffer strips around the body frequently is supported by the "underplacement" of a continuous tape, tacky on both sides (rubber based).

Subsupplied Parts. These parts are generally manufactured by subcontractors and are installed in the painted body in the course of trim line at the automobile assembly plant.

In the interior of the vehicle, prefabricated roofs consist of outer "decorative layers" that are bonded to the layers imparting mechanical rigidity (emulsions, contact adhesives, polyurethane hot-melt adhesives). Contact adhesives, polyurethane adhesives, and high-frequency welding auxiliaries (resin solutions or PVC-based emulsions) are used for attaching textiles or plastic films to the supports made of molded fiberboard or plastics for door inner panels. Emulsions, contact or polyurethane hot-melt adhesives are used for bonding textiles to molded fiberboard or plastics for parcel shelves.

On the exterior of the vehicles, rubber profiles around windows and doors and the trunk lid are "lap" bonded with cyanoacrylate adhesives because continuously extruded profiles cannot be used for sharp corners. To obtain effective sealing and better sliding in the case of sliding windows, these profiles are flocked by means of polyurethane-based flocking adhesives. The lenses of headlamps are joined to the housing by polyamide or polyurethane hot-melt adhesives and epoxy resins. The layers of laminated safety glass are bonded with a film based on poly(vinyl butyral).

Engine, Transmission, Chassis. In engines and transmission systems, components that must transmit torques, such as gear rims, are bonded with anaerobic adhesives (diacrylate based). The same group of products is used for thread locking. Gaskets formed in place in these areas consist of silicones or rubber – phenolic resin.

Rubber-to-metal bonding agents are used for the production of vibration-damping members of rubber and metal, developing their adhesion to the metal during vulcanization of the rubber. The same applies to the production of shaft seals. Clutch and brake linings are applied by means of nitrile rubber – phenolic resin adhesives.

In oil and air filters, the membranes are embedded in adhesives and/or fixed to metal frames (PVC plastisols, polyurethanes).

Future Trends. Short- and medium-term trends, such as the increasing use of plastic (sheet- or injection-molded compounds of glass-fiber-reinforced, unsaturated polyester resins, glass-fiber-reinforced polyurethanes, and polycarbonates) and lightweight alloys such as aluminum and magnesium will require new concepts for adhesives and sealants. New curing mechanisms are being sought to meet future production concepts (modular assembly). In addition the need to separate the multimaterial mix by disbonding processes for recycling and repair are special targets to be met by modern materials. Robots will guarantee further automation, lower baking temperatures will save energy, and the partial introduction of organic-coated steel panels for improved corrosion resistance will require slightly more versatility of the adhesives.

In the long term, materials for improved safety and comfort of vehicles of lightweight construction are required.

1.9.17. Adhesives in Aircraft Construction

The principle attractiveness of adhesive bonding in comparison to competing joining technologies (e.g., screws, riveting, brazing, welding) for the aircraft industry is mainly due to the following aspects:

- High aerodynamic surface quality of bonded components due to smooth and precise contours
- Superior properties with respect to stiffness, fatigue, and damage tolerance due to area-spreaded load transfer (no notches) with damping characteristics (vibration, sound)
- Nearly universal and simple joining of different materials without any weakening (minimal thermal exposure/stresses, surface damage)
- Simple realization of extremely lightweight sandwich and multilayer designs
- Gas- and liquid-tight joints (no crevice corrosion)

All these advantages can be utilized for saving weight or increasing life expectancy. Also reductions in production cost may be realized.

Disadvantages are:

- Strength properties are comparatively highly influenced by the environment (temperature, moisture, chemicals)
- Some processes require relatively high production standards
- Nondestructive testing is demanding and of limited significance
- Technology-specific destruction mechanisms exist (delamination, bondline corrosion)

However, many of the risks combined with the application of bonding can already be eliminated by means of careful and experienced design.

The application of adhesives in early aircraft manufacturing was not very different from the general evolution of this joining technology. For example, variations in the stability of wooden constructions against moisture due to less suitable adhesives such as casein were experienced. However, impressive parts like the wooden wing of the pre-World War II Fokker F-36 airliner (33 m wing span) were bonded with no in-service problems reported.

The gathering of specific know-how in the aircraft industry started with the transition to metal structures made of aluminum. This step was first made during World War II in the United Kingdom in the light bomber "Hornet". In this aircraft some structural aluminum parts were adhesively bonded with newly developed thermosetting phenolic adhesives instead of the cold-setting acid-hardening urea – formaldehyde type formerly applied with wooden components. This two-component system, a liquid phenolic resin plasticized with poly(vinyl formal) powder, has to be cured at 155 to 175 °C and quickly found its way into civil aircrafts like the DeHavilland "Dove" and the first jet airliner "Comet" [189]. During investigations into the crash series of the latter model, the bonding technology was suspected to be the reason for catastrophic structural failures. The investigations ended with a complete rehabilitation of the bonds, and the adhesive is still in use today in its original form or as a one-component precured (B-staged) film adhesive (e.g., skin panels of the Avro Regional Jet).

The most successful application example of this technology is the Fokker F 27/F50 "Friendship", in which about 70 % of the structure (ca. 550 parts) are bonded. The proven durability (> 1000 aircraft in service for up to 30 years) of this design is also due to the superior fatigue properties compared to a riveted construction. Additionally the chromic acid anodizing (CAA) of the aluminum surfaces before bonding resulted in corrosion-resistant joints.

Far less successful were further developments in the 1960s, when epoxy-based adhesives were introduced because of their higher temperature resistance and increased peel strength. Since no difference in mechanical properties were found between anodized and chromic sulfuric acid (CSA) etched aluminum, the advantage of low production costs was exploited. Delaminations of bonds after only two years in service and bondline corrosion events led to the recognition that the etch treatment produced an inconsistent oxide surface. Long-term salt-spray tests revealed the whole to be insufficiently stable to corrosive environments, especially as experienced in the bilge area of aircrafts. Expensive repairs and design modifications were the consequence.

The new standards then established [190], [191] included, besides the reintroduction of anodizing processes, careful edge protection, mandatory use of corrosion-inhibiting primer, and improved epoxy film adhesives with a curing temperature of at least 120 °C; this represents the technology mostly used up to today. Typical design rules require shear stresses of 10 MPa (static) and 1 MPa (dynamic). Any peel stresses should be avoided, but for safety purposes the bonds should bear about 4 N/mm in the floating roller peel test (e.g., EN 2243-2 [192]). In shear stress/shear strain tests (e.g., EN 2243-6 [193]) these adhesives show besides a nearly linear response up to at least 20 MPa and a failure strain of greater than 0.5.

Other lessons learned refer to the design of aluminum sandwich components consisting of a honeycomb core skinned with sheets. This extremely rigid and lightweight design is economically producible only by adhesive bonding, which also achieves superior stability against sonic fatigue. Durability against corrosion can only be attained if the parts remain completely airtight in service or, in case of acoustic panels around the engine, which are regularly heat dried during service, are well ventilated by perforations with access to all cells. Preventive anodizing and priming with corrosion-inhibiting primers also of the core material has not been applied very often till today due to the higher material costs. Progress in the bonded-sandwich design of the acoustic panels has made a substantial contribution to reduction in noise emissions of jet engines in the last two decades.

Besides the chemical industry, the aircraft industry also pioneered the utilization of titanium alloys and developed adhesive bonding as an option for joining workpieces. Similar to aluminum, the pretreatment process is decisive for successful and durable bonding. Simple etching with fluorine-containing acids resulted in bonds which are highly prone to moisture intrusion by diffusion and adhesive failure. Alkaline etching processes are a better alternative, and most durable bonds are obtained with anodizing procedures [194], [195]. The reliability of adhesively bonded titanium is indicated by the fact that the joints between the titanium fuselage fittings and the carbon fiber reinforced plastic (CFRP) wing skins of the US fighter aircraft F18 "Hornet" are adhesively bonded.

To a certain extent the aircraft industry is returning to its roots by applying more and more fiber-reinforced plastics, starting with G(glass)FRP, going to A(aramid)FRP, and now mostly CFRP, because these materials possess some similarities to wood. For all FRPs the load-spreading properties of adhesively bonded joints is a key issue, and successful adhesive bonding requires careful surface conditioning. Today mostly two processes are used: the peel ply technique and mechanical abrasion (sanding).

The peel ply technique integrates economical production of FRP parts with the advantage of obtaining reproducible surface conditions by using an outer woven ply (generally made of polyamide) during cure of the composite part. This ply, which is generally impregnated with an excess of the matrix resin of the composite part during the curing process, is removed (peeled off) just before the subsequent bonding process is started by applying the adhesive to the mating surfaces. A prerequisite for the successful application is that the peel plies are free of release agents and other contaminants.

More versatile are sanding processes. Especially the brushing of the surfaces with Scotchbrite pads and a slurry containing quartz particles and tensides cleans and prepares the substrates reliably. Prepregs can be directly bonded to such surfaces on cured laminates without using additional adhesive.

The workhorse adhesive materials for structural bonding in the aircraft industry are still the modified epoxy film adhesives curing at 120 °C or, for parts exposed to elevated temperatures (> 80 °C), 175 °C. Modification normally means the addition of other polymers that form a secondary phase of microscopic particles in the bondline and act as crack stoppers by distributing the stress of an advancing crack tip in the surrounding volume. For this purpose carboxy-terminated butyl nitrile (CTBN) rubbers are often used, but the addition of thermoplastics (e.g. polyamide, polysulfone) is also common. The epoxy resin itself is often made out of oxirane-terminated bisphenol A monomer and oligomer fractions. Additives are used to improve temperature and fire, smoke, toxicity (FST) properties. Curing agents are generally alkaline (amines, amides), often supported by accelerators, especially in the case of the adhesives curing at 120 °C.

Only for specific tasks with higher thermal stresses are polyimide-based adhesives utilized. Special products can withstand permanent in-service temperatures of 250 °C and more, but the application is economically often less attractive, so that such bonds are found mainly in military and aerospace components.

The most important room temperature (RT) curing two-component adhesives are also based on epoxy chemistry. These versatile materials are not only used for bonding, but also for other purposes like shimming, potting, encapsulation, and even impregnation of fibers and fabrics. Their domain is semistructural and interior joining. Because of early failures, their usage in structural applications (see above) has been restricted, but modern knowledge about failure mechanisms provides a basis for a return to the structural field. Recent developments have created materials with performance properties very similar to those of the tried-and-tested film adhesives. Especially the toughening properties were improved, and maximum service temperatures generally range from 80 to 175 °C. Hence these materials are today a powerful tool for economical repair and maintenance tasks.

1.9.18. Adhesives and Sealants in Electronics

Introduction. The use of adhesives in the electronics industry is experiencing a period of rapid growth at the moment. Adhesives are increasingly expected to perform a variety of tasks in electronics assembly, including conducting or insulating electricity, conducting heat, sealing, and protecting, besides the more usual functions of mechanically fixing components and assemblies.

Conductive Adhesives. *Isotropically electrically conductive adhesives* (ICAs) are widely used in the electronics industry when high-temperature soldering processes are unsuitable. Typical applications include silicon die attachment, component attachment in multi-chip modules (MCMs), surface mounted printed circuit board repair, and electromagnetic/radio frequency (EMI/RFI) shielding.

In all cases the conductive adhesive consists of high contents of conductive filler in an insulating polymer (adhesive) matrix. The choice of filler and adhesive matrix is dependent on the end use.

Die attachment adhesives are normally epoxy- or polyimide-based and contain 70 – 80 wt% silver flake. Silver flake is the common choice for conductive adhesives as the oxide layer formed on the flake is also conductive.

Polyimides became popular for die-bonding adhesives because they are cleaner (in terms of ionic contaminants) than equivalent epoxy products. The market at present, however, is dominated by high-purity epoxy adhesives. Polyimides are always applied from solvent solution and require higher curing temperatures than epoxies. They are stable to higher temperatures. Solvent-borne thermoplastics pastes (or cast-film preforms) have been used for some lower reliability die-attachment applications.

Epoxies form the basis of products in the multi-chip module (MCM) bonding and surface mount technology (SMT) repair areas. Two-pack room-temperature products (amine cure) are common in repair applications. Two-pack systems with acid anhydride curing are available as heat-cure products where a more robust final product is required. These two-pack systems are often pre-mixed by the vendor and supplied as a one-pack system requiring low-temperature storage. One-pack latent-heat-cure systems are also available based on dicyandiamide or imidazole adducts.

Electromagnetic/radio-frequency interference shielding materials have to meet much lower demands in terms of overall electrical conductivity (typically 4–5 orders of magnitude lower than a silver-flake-filled adhesive). This means that cheaper conductive fillers can be employed, for example, silver-coated copper flake, nickel flake, and carbon black. Typically the adhesive has to form a compliant joint between two mating surfaces, and hence room temperature vulcanizing or heat-cure silicone is often a convenient choice of matrix material.

Anisotropically conductive adhesives conduct in one direction only. They are often referred to as *z*-axis conductive adhesives. These materials have found widespread use in LCD, Liquid crystal display, assembly due to their ability to resolve small gaps between conductor tracks and the fact that solder is not suitable for joining on glass substrates.

The filler loading in anisotropically conductive adhesives is much lower than in conventional isotropic adhesives (typically 5–10 vol% as compared to > 30 vol% for ICAs). The fillers consist of spherical particles that are often monodisperse; examples include gold-plated polymers, solid gold, silver, and nickel. The filler loading is carefully chosen to insure that the probability of short circuits between adjacent conductors is low but that there is high probability of conduction along the *z*-axis. The adhesive matrix is usually epoxy-based and the most common product form is B-stagable film. These films are cured for short times at high temperature (20 s at 180 °C) under pressure.

Thermally conductive adhesives are commonly used as an interface between heat sinks and heat sources (e.g., high-power semiconductor devices). Their function is to give a mechanical integrity to the bond between the heat sink and heat source, but more importantly to eliminate air (which is a thermal insulator) from the interface area.

These adhesives consist of a polymerizable liquid matrix and large volume fractions of electrically insulating thermally conductive filler. Typical matrix materials are epoxies, silicones, urethanes, and acrylates, although solvent-based systems, hot-melt adhesives, and pressure-sensitive adhesive tapes are also available. Aluminum oxide, boron nitride, zinc oxide, and increasingly aluminum nitride are used as fillers for these types of adhesives. The filler loading can be as high as 70–80 wt%, and the fillers raise the thermal conductivity of the base matrix from 0.17–0.3 $W\,m^{-1}\,K^{-1}$ up to about 2 $W\,m^{-1}\,K^{-1}$.

Coating and Protection Products. *Conformal Coatings.* A conformal coating is a thin polymeric layer applied to a printed circuit board (PCB) by brushing, dipping, or spraying as the last processing step. The coating provides a barrier which protects the PCB (conductors, solder joints and components) from environmental attack during service life of the product. Not all of the circuit board is coated; areas such as connectors, switches and relays are covered with tapes, rubber caps, or removable masks to prevent the conformal coating impairing their operation. Conformal coatings are available in a number of different formulations; these include polyurethane (solvent and nonsolvent versions), silicones, acrylates, acrylated urethanes, and epoxies.

The choice of coating depends on the chemical and mechanical (abrasion) resistance required, the degree of rework envisaged, and the method of curing. Polyurethane coatings have excellent humidity, chemical, and abrasion resistance. However rework can be difficult. Silicones give the greatest protection during thermal excursions but they are unsuitable where abrasion resistance is required. Epoxies provide excellent chemical and mechanical resistance, but the inherent mechanical stiffness of the coating can be detrimental to stress-sensitive components. Rework is difficult. Acrylated urethanes offer a combination of fast cure and excellent solvent resistance.

Parylene is a completely different class of coating. The *p*-xylylene monomer is stable as a gas at low pressure but polymerizes spontaneously on any surface on which it condenses. The polymer formed has excellent moisture, chemical, mechanical, and electrical properties, but expensive application equipment is required.

Encapsulants encompass a broad range of materials whose primary function is to protect electronic components from detrimental chemical, mechanical, electrical, or thermal environments.

Encapsulants are usually characterized as either potting compounds, glob-tops, or underfills, depending on the end use. Potting compounds are materials that are used to protect final assemblies (relays, component terminals, electrical contact assemblies, etc.). Glob-tops and underfills are specifically used for silicon die protection.

Three different potting compounds are commonly used: silicones (for low-temperature applications), polyurethane (for applications up to 120 °C) and epoxies (for general applications). Potting compounds should provide good adhesion to the assembly materials/substrates, low cure shrinkage (and hence low internal stress), and in large sections, low heat release on curing.

Glob-tops offer protection to silicon die wire-bonded directly onto a substrate (PCB, polyimide flexible circuitry or ceramic). The liquid encapsulant is designed to flow between the wire bonds to form a protective coating over the integrated circuit. In some cases the pitch between wire bonds necessitates the use of low-viscosity liquids, and in these cases a dam of compatible adhesive is first dispensed around the IC to prevent the glob top spreading to other areas of the PCB. Typically, glob tops will have low levels of ionic contaminants and low coefficient of thermal expansion (CTE). The CTE is an important consideration in all encapsulant applications; the primary requirement is that the CTE of the encapsulant closely matches that of the silicon die, substrate, and wire bonds to minimize stress in the joint (particularly under conditions of thermal cycling).

Underfills are a specific class of adhesives designed to protect silicon dies which are soldered active face down onto the PCB. In these flip-chip applications, the underfill material flows beneath the die by capillary action. These materials are generally highly loaded with inorganic fillers to reduce the coefficient of thermal expansion.

Anhydride-cured epoxies give the best combination of properties for most glob-top and underfill applications, in particular their high glass transition temperatures are desirable. Due to the need for single-component materials (for ease of handling and bubble-free packaging) the anhydride and epoxy are premixed. The resultant one-pack system is stored at – 40 °C to give an acceptable product shelf life and a pot life of 8 h (or more) at room temperature. In some (lower reliability) glob-top applications UV curing acrylates or UV cationic epoxies have been used.

Surface Mounting Adhesives. Adhesives are commonly used to hold electronic components in position prior to wave soldering. One-component heat-curing epoxies are the most usual choice of surface mounting adhesive (SMA). These materials are thixotropic pastes based on bisphenol A or F type liquid epoxy resins and latent curing agents. The curing schedule of these materials is typically 3 min at 120 °C, with a six month (or longer) shelf life at 5 °C.

Acrylic SMAs form a smaller class of products. These materials are cured with both UV and heat, the UV curing mechanism insures rapid fixing, while the bulk of the adhesive (under the component) requires a heat-curing step. The shelf life of these products is long (12 months at ambient temperature) but overall product performance is inferior compared to the epoxy analogues.

Various adhesive application methods are available (dispensing, pin transfer and screen/stencil printing) and different adhesive rheologies are required in each case.

Adhesives for Liquid Crystal Display Manufacture. There are three main adhesive applications during the manufacture of a liquid crystal display (LCD): temporary plate fixing, main seal, and end seal. A number of LCD cells are made from single large flat glass plates. In the first process, the mainseal (or LCD gasket) is printed onto the glass to define individual cells, this is usually a solvent-borne epoxy. Once the solvent has been removed, a second glass plate is placed above the first and the mainseal is cured. To hold the two plates in the correct orientation while the mainseal is fully cured a temporary fixing adhesive is used. This is often a UV acrylic material. The final production step involves filling the cell with liquid crystal through a gap left in the mainseal and subsequently sealing the gap. This end-sealing operation is the only occasion when liquid adhesive comes into contact with the liquid crystal and hence the choice of adhesive is vitally important. Specially formulated UV acrylic formulations are most widely used for this application.

1.9.19. Medical Adhesives

Introduction. The use of adhesives in medical applications has a long history and may date back some 4000 years. Today, adhesives are used in a wide variety of surgical techniques albeit on a limited scale. The type of adhesive used depends on its application. It has to meet certain criteria with regard to its strength, toxicity, degradation pathway, and overall safety and efficacy. This applies equally to pressure-sensitive adhesives used in simple wound dressings to more advanced sealants that are used to plug small holes in blood vessels following cardiovascular surgery. Almost all branches of medicine have experimented with the use of adhesive to replace or augment traditional techniques. This section briefly reviews some of the more commonly used adhesives in medical and dental applications.

Tissue Bonding and Wound Sealing. Skin acts as a barrier to infection and dehydration. When it is breached in the form of a cut it is desirable to have it sealed quickly and effectively. Cyanoacrylates (CAs) have been developed as an alternative to the standard method of suturing (stitching) a wound [196]. CAs have the advantage over sutures that complete tissue closure can be effected, thereby preventing further fluid loss. Normally higher ester forms of CAs such as butyl are used. These result in less tissue inflammation than the methyl and ethyl esters, possibly due to slower degradation. As the polymerization process is exothermic, care must be taken to ensure that the setting speed is not too rapid, otherwise thermal damage to the surrounding tissue may occur. This is circumvented by using very small amounts of the adhesive and by using higher esters. The rate of polymerization is related to the size of the ester alkyl group. Generally, the larger the alkyl group, the longer the setting time, and hence there is more time for the heat generated to dissipate. Sterilization of CAs can be problematic since it often results in premature polymerization. Sterilization of CAs by dry heat and by gamma irradiation has now been described in the literature.

Trade Names. Indermil, Loctite Corporation and Histoacryl, B. Braun, Aesculap, B.B.D. (based on cyanacrylic butyl ester); Dermabond, Closure Medical Corporation, (based on cyanacrylic octyl ester).

Fibrin [197] is the body's natural glue or sealant and is produced in response to injury and bleeding. It helps form a clot, thereby sealing damaged blood vessels. Fibrin is formed by the cleavage of fibrinogen (a larger, soluble, blood-based protein) by the enzyme thrombin. This causes a structural modification resulting in the precipitation of fibrin to form a fine mesh which traps escaping blood cells. Fibrin acts well as a sealant and as a hemostatic agent, buts lacks the bond strength of synthetic chemical adhesives such as CAs. The individual components involved in this process can be purified from blood and be used in a variety of medical applications. It is especially useful in cardiovascular surgery.

Although they have quite different properties CAs and fibrin have been used for the same application. For example, both have been used to attach skin grafts to wound sites; to control bleeding and promote hemostasis; in vascular and intestinal anastomoses; and in ophthalmic surgery. Fibrin, being naturally derived, is more biocompatible. However, it is derived from blood and there is concern about transmission of pathogens. Fibrin(ogen) can be sterilized by a combination of either solvent/detergent or heat treatment to inactivate viruses, followed by ultrafiltration.

Trade Names. Tisseel, Immuno AG; Beriplast, Beringwerke.

Adhesives for Protective Dressings and Related Applications. Pressure-sensitive adhesives (PSAs) serve many applications in this area. Acrylates are the preferred elastomeric components of PSAs for general dressings. Acrylic hot-melt PSAs are useful for nonwoven web carriers due to their limited penetration. Specially chosen PSA formulations are used where allergic skin reactions are known or likely. Silicone-based PSAs may also be used in sensitive applications such as the attachment of prostheses and monitoring equipment probes. PSAs are also used to attach drug-delivery systems to the human body. Where the PSA forms a layer between the delivery source and the body tissue, particular care is needed to ensure adhesive function while facilitating drug availability. The design of such products depends largely on the size of the drug molecule.

Orthopedic Adhesives/Bone Cements. Acrylic bone cements [198] are the only group of materials currently used to anchor long-term implantable devices to the neighboring bone. Though the bone cements have drawbacks, including trauma created by a highly exothermic hardening process, over 90 % of hip and knee repairs function well for 15 years. The bone cements consist of separate powder and liquid components which are mixed carefully prior to application.

Though many products exist, the powder component always contains methacrylate polymer, a polymerization initiator, and a radiopaque medium. The liquid component contains methyl methacrylate with small amounts of accelerator for the hardening process. Hand mixing may be used. However, many proprietary mixing devices exist to give optimal properties and reproducibility.

Trade Names. Simplex, Howmedica; Palacos, Zimmer.

Dental Adhesives. Adhesive technology, cements and composites have added greatly to restorative dental surgery. Restoration of teeth with dental fillings is now often carried out with resin-based composites. These consist essentially of an organic methacrylate functional resin matrix, inorganic fillers, and coupling agents. Successive products have enhanced properties, in terms of the curing process, ease of use, and wear resistance. UY, Light-curing systems are now popular since they allow relatively unlimited working time followed by rapid hardening when exposed to light of the appropriate wavelength. The bonding of crown materials is usually done with similar compositions, though here systems which cure by mixing reactants are necessary since light only reaches peripheral areas.

Glass – ionomer cements have taken a major place in dental treatments as restorative filling materials and also in a range of more adhesive applications due to their ability to bond to both dentine and composite fillers. Acid-etching techniques are well established for the bonding of resins to enamel.

Trade Names. P 100 (Composite Dental Adhesive), 3M; Dyract (Glass Ionomer Cements), Dentsply.

1.9.20. Household Adhesives

Household adhesives are used for gluing, assembling, or repairing a variety of materials in the household. They include a large number of different adhesive types, usually offered in smaller packages such as tubes, bottles, or tins.

The following list provides a survey of adhesive types frequently used in households:

- Paper adhesives in solid (glue sticks), liquid, or cassette form (adhesive rollers)
- Multipurpose adhesives
- Adhesive tapes and pads
- Cyanocrylate adhesives
- Hot-melt adhesives
- Two-component adhesives
- Contact cements
- Plastic adhesives
- Wood glues
- Assembly adhesives

The term do-it-yourself adhesives has also become generally accepted besides the term household adhesives. A clear differentiation between the two ranges of adhesives is not possible. The above-mentioned household adhesives also include DIY adhesives. In addition to the above applications, DIY adhesives are also used for light renovation work in and around the house and are therefore also offered in larger packages or for activities such as wallpaper hanging. They include:

- Building chemicals such as tile adhesives, joint mortars, fillers, leveling compounds, and plasters
- Wallpaper pastes
- Flooring adhesives
- Sealants
- Polyurethane foams

Glue Sticks (Paper Adhesives, Solid). Glue sticks are used for gluing paper, cardboard, photos, and labels. Most of these sticks are solvent-free and contain a soap gel as builder and natural and/or synthetic polymers as adhesive component. They are also marketed as refillable sticks.

Of late, higher performance glue sticks based, for example, on polyurethane, which can be used to glue other materials such as wood, plastics, and metals have become available.

Trade Names. Pritt Stick (Henkel, Germany); UHU Stic (UHU, Germany); Tesa Klebestift (Beiersdorf, Germany); Fueki Stick (Fueki, Japan); Glue Stic (Avery Division, USA); Glue Stic (Grand & Toy, Canada); Kores Glue Stic (Kores, India); Glue Stick (Scotch 3M, France).

Paper Adhesives, Liquid. Liquid paper adhesives are used in the household for gluing paper and cardboard. These mostly solvent-free adhesives are partly based on natural raw materials such as dextrin, starch, and cellulose derivatives (pastes). They are sold in bottles and tubes but also as pens.

Trade Names. Pritt Pen (Henkel, Germany); UHU klebefix (UHU, Germany).

Adhesive Rollers (Paper Adhesives in Cassette Form). Adhesive rollers are newly developed adhesive systems for gluing paper, cardboard, and photos. The adhesive film is applied to a paper or foil carrier which is rolled up and placed in a cassette, usually with automatic take-up. This adhesive film is then rolled onto the parts to be glued on a solid surface. Adhesive rollers have the advantage over conventional paper adhesives that no drying time needs to be allowed after application of the solvent- and water-free adhesive film. Adhesive rollers are offered in various sizes, sometimes with refill cassettes. They are available for permanent and nonpermanent applications.

Trade Names. Pritt Roller (Henkel, Germany); Uhu multi roller (Uhu, Germany); Tombow (Japan).

Multipurpose adhesives are very versatile and can be used to bond most materials such as cardboard, paper, felt, textiles, wood, metal, glass, and plastics. They are also used for small-scale assembly or as a handicraft adhesive. Depending on the material, the bonding strengths achieved are different and in most cases on a medium level.

The solvent-free multipurpose adhesives contain transparent to opaque dispersions or solutions based on polyurethane or acrylate. In addition, white dispersions based on poly(vinyl acetate) or acrylate (so-called white glues) are available for all-purpose applications. The solvent-containing types mostly consist of poly(vinyl acetate), vinyl acetate copolymers, or nitrocellulose, dissolved in a solvent mixture of esters, ketones, and alcohols.

Trade Names. Pattex Haushaltskleber transparent (Henkel, Germany); Pritt Alleskleber (Henkel, Germany); Uhu flinke Flasche (Uhu, Germany); Tesa Vielzweckkleber (Beiersdorf, Germany); Elmer's school glue (Borden, USA).

Adhesive Tapes and Pads. Adhesive tapes make use of foils, paper, or fabric as a carrier material which is coated with a permanently sticky film. In the household these tapes, which can be bought in different types and sizes (one-sided or two-sided adhesive tape), are primarily used for packaging, for small-scale assembly, and for masking when doing renovation work.

By contrast, two-sided adhesive pads consist of foamed carrier materials or extruded caoutchouc pastes and are used to fix lightweight objects on to smooth, level surfaces.

Adhesive pads for easily removable bonds that do not require a carrier material and are particularly suited for fixing posters, graphics, and other lightweight objects, even to sensitive substrates such as wallpaper.

Trade Names. Pritt Klebepads (Henkel, Germany); Tesa Klebebänder (Beiersdorf, Germany); Tesa Power Strips (Beiersdorf, Germany); Duck Tapes (Manco, USA); Scotch (3M, USA).

Cyanoacrylate adhesives are used in households for repair work. As these adhesives cure very fast and hardly bridge larger cracks, they are only suitable for small, snug fitting parts. Cyanoacrylate adhesives can be used to bond almost any material, both porous and smooth.

The main ingredient of this adhesive is a cyanoacrylic acid ester (methyl, ethyl, or butyl). Special esters such as methoxyethyl esters are employed for the odorless types. Cyanoacrylate adhesives are mainly available in small aluminum tubes or plastic bottles at different viscosities ranging from highly liquid to gel-like.

Trade Names. Pattex Blitz-Kleber (Henkel, Germany); Super Attak (Loctite, USA); Uhu Sekundenkleber (Uhu, Germany); Krazy Glue (Toagosei, Japan); Aron Alpha (Toagosei, Japan); Cyanolit (3M, USA).

Hot-melt Adhesives. In household applications, hot-melt adhesives are used in cartridges together with the necessary gluing pistols. They are suited for repair and small-scale assembly work, being able to bond almost any material. The special advantage of hot-melt adhesives is that the bonded parts already can be subjected to stress or further processed after a few minutes. A certain drawback is their high processing temperature of ca. 200 °C, which renders the bonding of temperature-sensitive materials such as styrofoam or plasticized PVC more difficult. Lately, so-called low-melt cartridges have come in use which can be processed at a temperature as low as ca. 110 °C. Furthermore, mini-pistols are offered which accommodate smaller cartridges, colored cartridges, or cartridges with a glitter effect.

Those hot-melts that find application in the household and DIY applications are usually based on ethylene–vinyl acetate copolymers (EVA) and contain tackifying resins (e.g. hydrocarbon resins). Today's pistols are equipped almost without exception with a mechanical feed, differ in their heating power, and are partly designed for cordless work.

Trade Names. Pattex Patronen (Henkel, Germany); Uhu Klebepatronen (Uhu, Germany).

Two-component adhesives that are used for household applications differ quite strongly as to their properties. For instance, there are epoxy resin adhesives that are very well suited for bonding glass, china, ceramics and metals. They set glass-clear, are dishwasher-safe, and therefore exceptionally well suited for repairing dishes or similar household articles. Usually, the two components resin (epoxy) and hardener (amine or mercaptan) are filled in aluminum tubes or plastic double syringes and are mixed before application according to the directions for use. Other two-component adhesives are based on methacrylate and contain a powder hardener (dibenzoyl peroxide in gypsum). These adhesives bond materials such as stone, plastics, ceramics, and metal;

they achieve very high strength and bridge cracks and are therefore well suited for repair and small-scale assembly work.

Further two-component adhesives are available based on polyurethane; here, the resin component contains a polyol (filled or unfilled) and the hardener component a polyisocyanate. These adhesives, too, are suitable for bonding metal, plastics, ceramics, and wood. Special highly filled types can be used both for bonding and as a filler; after curing they can be processed like wood.

Two-component adhesives are used in the household when more difficult materials such as metals, glass, or certain plastics are to be bonded, when high strength is required, or when larger gaps must be bridged.

Trade Names. Pattex Stabilit Express (Henkel, Germany); Pattex Kraft-Mix (Henkel, Germany); Uhu plus (Uhu, Germany); Scotch Weld (3M, USA); Bond E (Japan).

Contact Cements. In the household, contact adhesives are used in many areas for gluing and repairing. They are mainly used when larger and heterogeneous items (e.g., wooden materials with decorative laminates) are to be joined, when difficult parts (e.g., veneer strips on the curves and edges of furniture) are to be fixed, or when flexible materials such as rubber or leather (e.g., shoe repair) must be bonded or repaired. Furthermore, contact adhesives can be used to bond materials such as cork, felt, plastics, soft foams, and metals.

Most contact adhesives contain a solvent and have a formulation that combines a base of synthetic rubber such as polychloroprene or polystyrene – butadiene with reactive phenolic resins and metal oxides. In addition, there are also transparent types based on polyurethane which provide excellent bonding results for soft plastics such as plasticized PVC (used in many household articles). Recently, a solvent-free generation of contact adhesives, e.g., based on acrylate, has entered the market. This type can be used to bond solvent-sensitive materials such as polystyrene foam.

Contact adhesives are sold either in cans or tubes with different viscosities (liquid or gel-like). Their application differs from that of other adhesive systems in a few points. The adhesive is applied to both parts to be joined, after which the solvent or water is allowed to evaporate (5 – 40 min, depending on the adhesive). After the evaporation time, bonding is possible within a defined period (open time) by pressing both parts briefly but with the highest possible pressure together.

Trade Names. Pattex Kontaktkleber (Henkel, Germany); Uhu greenit (Uhu, Germany); Elmer's (Borden, USA), Cemendine (Japan).

Plastic adhesives are used in the household or for model-making. They are able to bond nearly all plastics, for example, polystyrene, ABS, plexiglass, polycarbonate, and rigid PVC. Plastic adhesives partly belong to the solvent-welding adhesives. As they contain solvents such as butyl acetate, the surface to be bonded is slightly solvated, and this results in a solvent-welding effect. Plastic adhesives are mostly offered in small bottles, partly equipped with fine-dosing nozzles.

Trade Names. Pattex Plastic (Henkel, Germany); Uhu Plast (Uhu, Germany).

Wood glues are generally ready-for-use, water-based white glues based on poly(vinyl acetate). In the household they are used for assembling or for repairing wooden materials. After applying the glue, the wooden parts to be joined must be cold- or hot-pressed for a certain amount of time. In addition to the conventional wood glues, the market offers special fast-curing and waterproof types, also based on reactive polyurethane.

Trade Names. Ponal Holzleime (Henkel, Germany); Uhu coll (Uhu, Germany); Dorus Holzleim (Dorus, Germany); Elmer's wood glue (Borden, USA), Rakoll (Fuller, USA).

Assembly adhesives are relatively new adhesive systems which also find application in households. Assembly adhesives are able to produce very high bonding strength and high initial tack so that the parts to be joined need not be fixed. They are used whenever the powerful fastening of mostly different materials is required (both small and large areas), i.e., in cases where nailing, screwing and drilling were necessary up to now.

Typical applications include for example the bonding of wooden panels, skirting boards, decorative and insulating panels, plasterboards, tiles, etc. Moreover, it is also possible to bond difficult parts such as mirrors onto absorbent or nonabsorbent substrates, styrofoam sheets on metal, or nonferrous metals.

Assembly adhesives are either available as dispersions or as one-component reaction adhesives in cartridges or tubes.

Trade Names. Pattex Montageklebstoffe (Henkel, Germany); Mastic colle neoprene (Rubson, France); One for all (Bison, The Netherlands); PU-Fixkleber (Compacta, Germany); Sikaflex 221 (Sika, Switzerland); Montagekleber MK (Fischer, Germany); Fix all (Soudal, Benelux); Draft basters (Manco, USA); Multifix (Quilosa, Spain).

1.9.21. Applications of Anaerobic Adhesives.

Applications of anaerobic adhesives [199] are as follows:

- *Thread locking*: The first applications for anaerobic adhesives were for locking threaded fasteners against vibration. Filling the "inner space" between a nut and bolt with a hard, dense material prevents self-loosening [200].
- *Thread sealing*: The effect of filling the space between threaded parts or the space between inner and outer pipe threads provides a seal which can prevent the leakage of oil and other fluids in machinery as well as prevent corrosion of the threaded parts.

Table 1. Regional distribution of the world adhesives market

Region	10^6 $	1000 t
Canada and Mexico	413	187
United States	5799	2277
South America	302	130
Western Europe	3733	1470
Eastern Europe	887	323
Middle East	314	114
Africa	230	86
Asia/Pacific	1882	654
Japan	1925	585
Australia, New Zealand, Oceania	175	65
World total	15 660	5891

- *Retaining*: Cylindrical press-fits and bearing assemblies can be retained with anaerobic adhesives, allowing accurate alignment and relaxed tolerances. Retaining and sealing of cup plugs and oil seals in castings is a major application.
- *Impregnation*: Powdered metal parts, porous castings, and welds can be sealed against leakage of liquids or gases. This impregnation can also allow such parts to be plated and improves their machinability.
- *Preapplied Films*: Thread lockers and sealants can be coated on threaded parts in the form of a dry-to-the-touch film. These preapplied materials remain inert until assembly releases a quick curing resin or an initiator.
- *Gasketing*: Anaerobic flange sealants can be applied manually or by automated methods such as tracing, stenciling, and screen printing. These products can replace a variety of preformed, precut gaskets, and they can also be used as a gasket dressings.
- *Structural Bonding*: Tough structural bonds can be achieved with some anaerobic adhesives for bonding components such as ferrite magnets, honing stones, identification tags, and decorative inserts.

1.10. Economic Aspects

More than 1500 adhesives manufacturers worldwide offer a variety of not less than 250 000 products. The worldwide turnover of the adhesives industry is believed to be approximately $ 16×10^9 corresponding to a total consumption of 6×10^6 t. These figures do not include cement-based products and urea and phenolic resins used as binders for the production of chipboard. Market studies and national industry federation reports are available for the United States, Canada and Mexico, Latin America, western Europe, eastern Europe, Asia/Pacific, Japan, and Australia. The regional distribution of the world adhesives market is given in Table 1.

Adhesives have a very wide spectrum of applications. They are applied in almost every industrial field. The largest markets are found in the paper, packaging, and board industry (35 %), the building, construction, civil engineering and craftsmen industry (28 %), and the woodworking industry (15 %). The forecasts for the future growth rates vary. The highest growth rates are expected in the automotive industry and the packaging and the converting sectors.

Considering the product groups, the highest growth rates are expected for hot-melt systems and polymer dispersions and emulsions. A continuing decrease will occur for adhesives based on natural polymers (animal and vegetable), solvent-based adhesives, and systems based on water-soluble polymers.

2. Adhesive Tapes

WERNER KARMANN, tesa Division, Beiersdorf AG, Hamburg, Federal Republic of Germany
ANDREAS B. KUMMER, medical Division, Beiersdorf AG, Hamburg, Federal Republic of Germany

2.1. Introduction

The first pressure sensitive adhesives were based on natural rubber and rosin. They were applied as patches and tapes for surgical use [203]. In the 19th century some early technical tapes were derived therefrom (e.g., Citoplast, P. Beiersdorf & Co., Germany), but the application range was very narrow. When spray painting was introduced in the automotive industry, such tapes were employed to mask color lines. They were imperfect and had to be improved [204]. This was the starting point for the development of industrial tapes.

Currently there are hundreds of varieties of pressure-sensitive tapes that differ in the adhesive base, the carrier material, and construction. They are used in practically all industries, in house and office, in hospitals and personal care.

Adhesive tapes generally consist of a flexible carrier or backing with a layer of a permanently tacky adhesive which sticks to most substrates by contact under pressure. There is no need for solvent or heat. The carrier often executes the function of the tape, e.g., to close a box, to protect or decorate a surface, to fill a gap, or to cover a wound.

The main types of adhesive tape are:

- *Pressure-sensitive tapes* consist of a carrier, coated, with adhesive on one side, and wound into a roll. The back side is often coated with a release agent to control the unwind force.
- *Double-sided tapes* generally consist of a carrier with adhesives on both sides and an intermediate release liner, to prevent contact of the adhesive layers when the tape is wound into a roll. There are also transfer tapes without carrier and tapes without release liner.
- *Adhesive pads* are pieces of thick layers of adhesives, or polymer foams with adhesive layers on both sides. They are often stretchable for ease of release from the surface.
- *Medical tapes* are composed of a backing coated with adhesive (porous, permeable or occlusive). Some have a wound pad.
- *Labelstock* consists of a writable or printable carrier coated with the adhesive on one side, which generally is covered by a release liner. This type is not treated here.

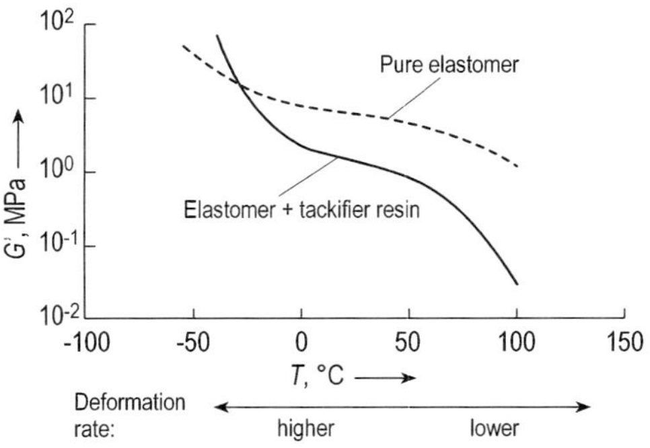

Figure 11. Effect of tackifier resin on the storage modulus of an elastomer [209]

2.2. Pressure-Sensitive Adhesion

Normally adhesives are applied in a liquid state to the substrate, wet the surface, and solidify to generate a bond, for example by evaporation of a solvent, cooling of a hot melt, or formation of a cross-linked polymer. Pressure-sensitive adhesives do not change their physical state, but they also must wet the surface of the substrate like a liquid and sustain loads like a solid. There have been many discussions and investigations concerning how they fulfill these conflicting demands.

The first theories were based on the observation of two phases at the surface of rubber – resin adhesives. One, having high resin content, was assumed to act as sticky spots that adhere to the surface, while the other, with a low resin content, gave cohesion to the system [205], [206]. But some contradictions remained unresolved [207].

Later investigations showed that combinations of rubbers with very effective tackifiers do not show the two-phase structure. Therefore, phase separation cannot be essential to pressure-sensitive adhesion. The above-mentioned conflict rather is resolved by the dependence of mechanical strength on the rate of deformation, that is, viscoelasticity. The tackifier resins decrease the modulus at the slow rate at which wetting occurs, and increase it at the higher rate at which the adhesion tests are carried out [208]. This is demonstrated in Figure 11, in which the mechanical strength is characterized by the storage modulus G' (the explanation is given later).

To achieve a good compromise between wetting the surface (adhesion) and bearing a load (cohesion) the modulus should be in the range of 0.1 to 1 MPa "in the time frame of the bonding process", that is 1 – 100 s (Dahlquist criterion [211], p. 101).

According to Williams – Landel – Ferry (WLF) principle longer time is equivalent to higher temperature. Thus at higher rates the modulus is shifted to the direction of lower temperature and vice versa [210].

2.2.1. Mechanism of Pressure-Sensitive Adhesion

Pressure-sensitive adhesion is a complex phenomenon determined by a number of physical processes. A comprehensive overview is given in [211].

Pressure-sensitive adhesion is determined by two key factors. First, the surface must be contacted to such a degree that molecular attraction forces come into action (wetting). The second factor is the energy of deformation of the viscoelastic adhesive until rupture or separation from the surface occurs.

The degree of wetting is controlled by thermodynamic and kinetic factors. The driving force for wetting a substrate is dependent on the surface energies of the adhesive and the substrate. The Zisman concept of critical surface tension provides the basis for estimating the surface energy of solids [212]. Typical values of critical surface tension (in mN/m) for polymer substrates follow [216]:

Polytetrafluorethylene	19.5
Poly(dimethylsiloxane)	22
Polypropylene	29
Polyethylene	34.5
Poly(methyl methacrylate)	41
Poly(vinyl chloride)	41
Polystyrene	42
Poly(ethylene terephthalate)	42.5
Polyamide 66	44

The influence of critical surface tension on peel adhesion, tack, and residues was investigated experimentally for rubber – resin [213], [214] and polyacrylate [215] pressure-sensitive adhesives (PSAs).

The critical surface tension of a PSA cannot be determined directly, for when the test droplets of low polarity liquids are dissolved in the adhesive the contact angle is not defined. Therefore, this value was derived from those of chemically similar polymers, resulting in 33 mN/m for polyisobutylene and 35 – 36 mN/m for standard acrylic and rubber – rosin PSAs.

The study showed a maximum of peel force and residues on substrates having a somewhat higher surface tension than the adhesive. Hence, to have good adhesion, the adhesive should have low surface energy or the substrates should be treated to increase their critical surface tension.

Adhesion is enhanced further by contact time and contact pressure. Both increase the area of the wetted spots kinetically by flow of the adhesive. This is especially important for rough surfaces and cross-linked adhesives.

The role of the deformation energy can best be demonstrated by peeling a pressure-sensitive tape from a rigid substrate. During separation typically the adhesive becomes filamented and the "legs" are stretched to such degree that they break or separate from the surface. The former is a cohesive and the latter an adhesive failure. The energy required to achieve this is the peel force.

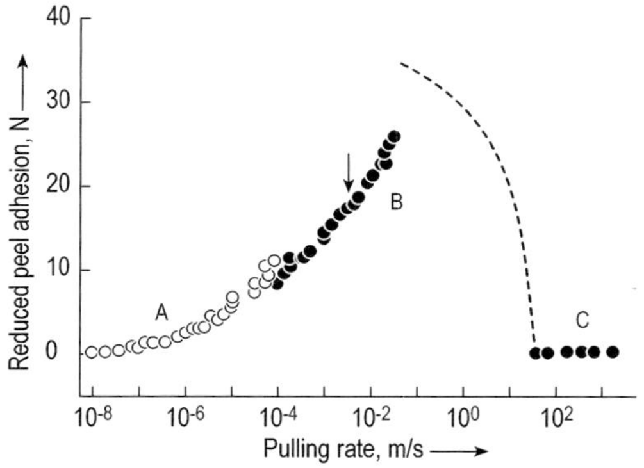

Figure 12. Master curve for the force of a tape peeled at 90 ° from a glass surface (transposed to 23 °C according to WFL principle [210]) ○ Cohesive failure; ● Adhesive failure [PET film with adhesive (60 % rubber, 40 % β-pinene resin)] [217]

The correlation between peel adhesion and the viscoelastic properties of the adhesive is reviewed in [211, p. 61 ff]. A comprehensive experimental analysis can be found in [217]. The peel force was measured at various pulling rates and temperatures and superimposed on a master curve. The results are shown in Figure 12.

At low rates (i.e., at high temperatures) there is a region A of cohesive failure with a clear influence of pulling rate.

In region B at higher rate (i.e., lower temperature) the adhesive is still legging, but separation occurs at the surface. In this region the magnitude of the peel force is influenced by the area of molecular contact (i.e., by wetting). At still higher rates is region C, where the adhesive is pulled from the backing. Here the peeling force is dramaticly lower and independent of the pulling rate.

Between B and C is a transitional region, in which the peel force jumps from a higher to a lower level, so-called slip-stick. The upper limit is shown by the dashed line. It is believed that this is a periodical change from rubbery to glassy response during peeling.

Typically, pressure-sensitive tapes are applied in region B. The effective pulling rate of the standard peel test at 180° is 2.5×10^{-3} m/s, which is in the middle of region B (indicated by an arrow in Fig. 12). At very low temperature, region C may be reached. For application under freezer conditions, special tapes with lower T_g are available.

To avoid cohesive failure at higher temperature the adhesive can be cross-linked. This shifts region A to lower rates (i.e., higher temperature) or suppresses it entirely.

Viscoelastic properties are influenced essentially by the glass transition temperature T_g. The optimal T_g is about 40 – 70 °C below the application temperature of the PSA. The T_g can be adjusted by means of the composition of the adhesive, especially by the concentration and the softening temperature of the resins.

The viscoelasticity and surface tension of adhesives are physical properties that are independent of the specific chemical composition. Therefore, the relations discussed for rubber-based adhesives are also valid for polyacrylates [218] and other elastomeric bases.

2.2.2. Dynamic Mechanical Analysis

The viscoelastic properties of a polymeric material can be described by its reversible and irreversible responses to deformation. These can be identified most easily by dynamic mechanical analysis (DMA). Usually, the adhesive is placed between two parallel plates, one of which is oscillating sinusoidally, and the torque is measured. From the amplitude and the phase shift of the sinusoidal stress – strain curve, the elastic component, which is in phase, and the viscous component, which is 90° out of phase, can be derived [211, p. 158 ff].

Thus, the viscoelastic properties are characterized by the storage modulus G' (the elastic component of the modulus), the loss modulus G'' (the viscous component of the modulus), and $\tan\delta = G''/G'$ (the tangent of the phase shift between stress and strain), which are functions of the deformation rate and temperature. The method offers valuable information for the development and the application of adhesive tapes.

The maximum of $\tan\delta$ as a function of temperature is generally identified as the (dynamically defined) glass transition temperature T_g, which is dependent on the deformation rate. Usually it changes by 6 – 7 °C per decade of frequency [219, p. 253].

A mixture of two compatible polymers with different T_g values exhibits a single T_g which is influenced by the T_g values and weight fractions of the two components. In many cases this T_g can be calculated approximately by the Fox equation (Eq. 1) [219, p. 277 ff]

$$1/T_{g_m} = w_1/T_{g_1} + w_2/T_{g_2} \qquad (1)$$

where T_{g_m}, T_{g_1}, T_{g_2} are the glass transition temperatures of the mixture and of components 1 and 2, and w_1, w_2 are the weight fractions of components 1 and 2.

A mixture of partially compatible polymers exhibits the T_g of at least one pure component and a T_g, which is dependent on the mixture, according to Equation (1).

2.3. Test Methods

Standard test methods are applied primarily for characterizing adhesive tapes and assuring product quality. To check whether a tape is suitable for a specific application generally requires more specific tests, which may be agreed with the customer.

Standard test methods for pressure-sensitive tapes are published by governmental and commercial organizations:

USA	Pressure Sensitive Tape Council (PSTC) [220]
USA	American Society for Testing Materials (ASTM) [221]
Europe	Association des Fabricants Europeens de Rubans Auto-adhesifs (A.F.E.R.A.) [222]
Europe	European Committee for Standardization (CEN) [223]

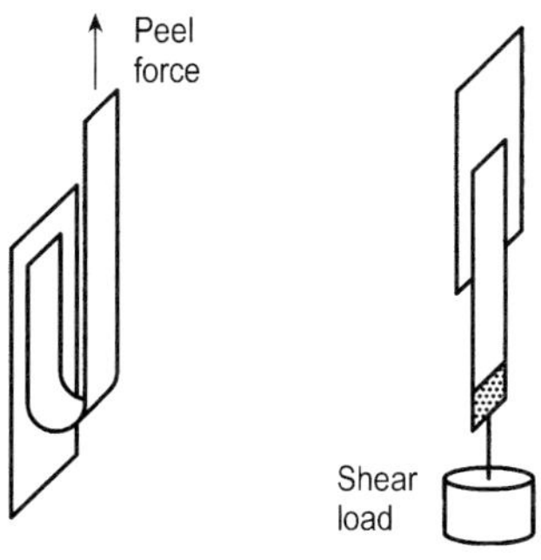

Figure 13. Standard test methods for peel and shear adhesion of pressure-sensitive adhesive tapes

Adhesive properties of pressure-sensitive tapes are characterized primarily by three parameters: peel adhesion, shear adhesion and tack.

Peel adhesion is a highly characteristic property of a PSA. It gives the force per unit width required to remove a PSA tape by peeling from a test panel under defined separation speed and angle (0.3 m/min and generally 180° (Fig. 13), sometimes 90°). The backing can influence the results by its stiffness or stretchability. A more detailed discussion is given in [211, p. 61 ff].

Shear Adhesion (Creep). The most significant example of creep is a tape gliding down under shearing load. Therefore, most commonly creep is characterized by the time a tape holds under defined shear load (Fig. 13). This often is called holding power or shear adhesion. Further examples of creep are the telescoping of rolls of stretchable pressure-sensitive tapes and the greasy spots occurring on a paper sheet mended by a transparent adhesive tape.

Resistance to long-term creep is critical for most applications of pressure sensitive products. According to WLF principle [210] long time is equivalent to high temperature.

The creep of a PSA can be minimized by increasing the molecular mass of the adhesive and by cross-linking it. But there are limitations, because some creep is necessary to wet the surface and to relax stress concentrations when the adhesive is stressed. This often is referred to as the balance between adhesion and cohesion. The relevance of creep to PSAs is discussed in depth in [211, p. 97 ff].

To understand the reason for a failure caused by creep it is important to identify whether it is a cohesive failure (nearly the same amount of adhesive on both sides) or an adhesive failure (residue on one side only). In the first case, the adhesive is too soft or undercured, in the second it may be too hard or overcured. Furthermore, there may be a weak boundary layer on one surface. Sometimes there are structured patterns of adhesive on the surface. This is an indication of fracture as a result of stress concentrations [211, p. 115 ff].

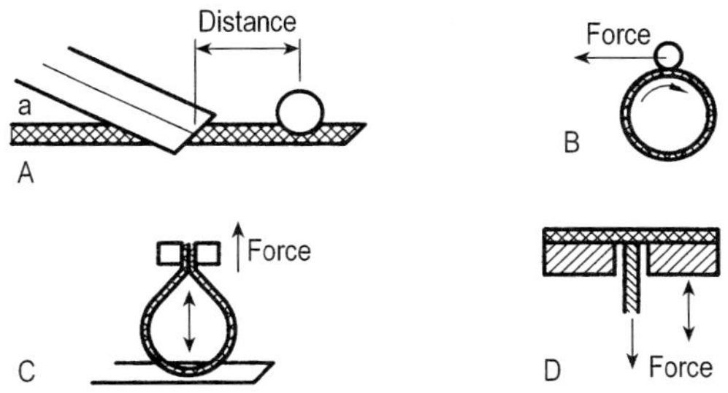

Figure 14. Standard test methods for the tack of pressure-sensitive adhesive tapes
A) Rolling-ball tack; B) Rotating-drum tack; C) Loop-tack; D) Probe-tack

Tack is the ability of a PSA to create a bond spontaneously under no or minimal external contact pressure. There are several aspects of tack, derived from different applications. Some consider the shortest time of contact, others the lowest contact pressure; some consider smooth, others rough surfaces.

The test methods can be classified into three groups (Fig. 14) [211, p. 38 ff]:

1) Rolling ball and rotating drum tests, in which the substrate runs on the surface of the adhesive. The measured quantity is the distance traveled or the imposed force, respectively.
2) Peel tests involving low contact pressure and short application time (e.g., the loop test).
3) Probe test, in which the tip of a probe is brought into contact under low pressure for a short time. The probe is withdrawn at a defined speed, and the butt tensile strength is measured.

Shear adhesion failure temperature (SAFT) gives information on the resistance of the PSA to stress at elevated temperature. The geometry is identical to the shear adhesion test (Fig. 13). The test is performed in a forced-air oven with temperature rising at constant rate (0.4 °C/min). The SAFT is the temperature at which the bond fails and the load falls down. As this method is not standardized yet, there are some differences in overlap and load between companies [225], [226].

Further tests on adhesive and physical properties of tapes are:

- The unwind force of pressure-sensitive tapes
- The influence of aging on the unwind force of tapes
- Stains left on a substrate after removing the tape
- Strength and elongation of tapes under load
- Water and solvent resistance

For electrical tapes there are a number of very specific tests. More information is given in Chapter 2.6.1.6.

2.4. Raw Materials

The key component of PSA is the elastomer. It has the main impact on the physical and chemical properties of the adhesive. Its glass transition temperature can be raised either by adding sufficiently compatible resins or by copolymerization with suitable monomers. The composition of the adhesive also influences its surface energy, which is important for the ability to wet a surface.

2.4.1. Elastomers

Polydiene Rubbers. Most important of these is natural rubber (NR). Additionally used are butadiene – styrene rubbers (SBR) and synthetic polyisoprene rubbers (IR). They have very high molecular masses (several 10^5 to some 10^6 g/mol) and low T_g (– 70 to – 40 °C).

NR is classified internationally according to its production process, purity and Mooney viscosity [227, pp. 12 – 17]. From this, information about the molecular mass of the adhesive can be derived, which is further influenced by the mastication and mixing process.

To achieve high tack and peel adhesion the T_g of the rubber must be raised by adding tackifier resins. The ratio of resins to rubber generally covers the range of 0.75 – 1.2.

Due to their main chain double bond, polydiene rubbers are prone to oxidative degradation and can be cross-linked by reactive resins and sulfur compounds [227, pp. 233 – 261]. Cross-linking can be achieved further by an electron beam (EB) and in presence of photoinitiators by UV radiation [228].

The adhesives are applied mainly from solvents, some are applied solventless by calendering or as water-based latices. Quite recently solvent-free hot-melt processes have been introduced [229].

Styrenic block copolymers are special polydiene elastomers (polyisoprene or polybutadiene) of moderate molecular mass (7×10^4 to 2.5×10^5) with end blocks of polystyrene [$(7 - 25) \times 10^3$ g/mol]. They are thermoplastic elastomers: At low temperature the polystyrene blocks form spherical hard domains that act as cross-links in the adhesive and result in very high cohesion. Above about 90 °C they soften to give a melt of comparatively low viscosity like a thermoplastic [211, pp. 317 – 373].

There are different types of block copolymers: linear ABA, AB, and branched $(AB)_n$ with various ratios of A to B (A is polystyrene, and B is polybutadiene, polyisoprene, or their hydrogenated derivatives).

They must be compounded with resins to achieve high tack and peel adhesion. All ingredients must be selected carefully so as to independently control their influence on the elastomeric and the styrenic blocks [230].

Table 2. Important monomers for acrylic PSAs

Monomer	T_g*, °C	Polarity
Soft monomers		
n-Butyl acrylate	−54	low
2-Ethylhexyl acrylate	−70	very low
Hard monomers		
Methylacrylate	+10	medium
Vinylacetate	+30	medium
tert-Butylacrylate	+43	low
Styrene	+100	low
Polar monomers		
Acrylic acid	+106	very high

* Quasistatic glass transition temperature of the homopolymer [239]

Block copolymers with hydrogenated elastomer blocks are much more stable than those with unsaturated ones but they have somewhat limited adhesive properties. New developments will overcome this limitation [231].

After coating, the hot-melt adhesives can be cross-linked to compensate for softening at higher temperatures [232], [233].

Block copolymers are especially useful for hot melt and high-solids PSAs.

Acrylics. Polyacrylates are copolymerized from different monomeric acrylates, optionally with some vinyl compounds. By selecting suitable comonomers, the T_g and the surface energy can be adjusted over a wide range. The most important monomers are listed in Table 2.

Further important parameters are the average molecular mass and its distribution, as well as the sequence of the monomers in the chain. These are set by the polymerization process.

Acrylic copolymers with suitably adjusted T_g are inherent pressure-sensitive adhesives, but they can be compounded with reasonably compatible resins to adjust tack, peel, and creep properties.

For most applications acrylic copolymers must be cross-linked. This can be performed by chemical reactions of the functional groups of the comonomers (amine with acid, epoxy with acid, diisocyanate with acid or hydroxyl, metal chelate with acid [234]) or by radicals generated by thermal decomposition of peroxides or by UV or electron beam (EB) radiation [235].

Most often acrylic adhesives are applied from solvent or as water-based emulsions. For special applications, like thick double-sided tapes, a solventless process has been developed in which a partially prepolymerized mixture of acrylic monomers is coated onto a liner and subsequently polymerized by UV radiation to form the adhesive [236]. Moreover, acrylic hot-melt adhesives cross-linkable by UV and EB radiation are being developed [237], [238].

Polyisobutylene and Butyl Rubber. Polyisobutylenes are polymers of low polarity and low T_g and are available in a wide range of molecular masses. For pressure-sensitive adhesives, polymers with a viscosity average molecular mass M_v of at least 10^6 are used in combination with low molecular mass types.

Butyl rubbers are copolymers of isobutylene with minor amounts of a diene that enable chemical cross-linking via unsaturated sites. The M_v is in the range of $(3-4) \times 10^5$. To improve the possibilities for cross-linking, it can be mixed with halobutyl rubber, which contains the halogen in the α-position to the double bond.

Polyisobutylene rubbers can be compounded with resins to raise the T_g and hence peel adhesion.

Ethylene – Vinyl Acetate Copolymers (EVAc). The majority of EVAc is used for hard tack-free hot-melt adhesives. Pressure-sensitive properties can be achieved with types containing 40 – 60 % vinyl acetate compounded with compatible resins. EVAc-based PSAs are used only for specialty tapes.

Silicone Elastomers. The main component of pressure-sensitive silicone adhesives is a silicone elastomer with a molecular mass of several 10^5 g/mol. There are two types: One is neat poly(dimethylsiloxane) having very low T_g (– 125 °C) [240] and critical surface tension. The other is modified by substitution of 5 – 15 % of the methyl groups by phenyl groups. This results in higher T_g (ca. – 86 °C), higher surface tension, and improved heat stability [241].

High peel adhesion is achieved by compounding with silicone resins. Since they are only partly compatible, this mixture must be stabilized by reaction of some free silanol groups in both components.

To increase cohesion, silicone elastomers can easily be cross-linked via radicals generated by thermal decomposition of peroxides or by electron beam radiation.

Silicone elastomer based adhesives cover wide range of service temperatures (– 70 to ca. 250 °C) and provide excellent aging resistance. Owing to their low surface energy, they are the only adhesives applicable to low-energy substrates like silicone release liners and PTFE.

Normally, the adhesives are applied from aromatic solvents. Newer developments use the noble metal catalyzed addition reaction of Si–H to $H_2C=CH–$ for high-solids and solventless silicone PSAs.

2.4.2. Tackifier Resins

Tackifier resins are generally brittle amorphous materials with high softening temperatures (typically 40 – 150 °C) [242], [243] and low molecular masses (some 10^2 to some 10^3 g/mol). Some are liquids that soften below ambient temperature. All should exhibit reasonable compatibility with the elastomer, which can be detected by the clearness of the mixtures or by dynamic mechanical analysis (see Chap. 2.2.2.). Compatibility can be predicted from the cloud point temperature in suitable mixtures of solvents [244].

The presence of high molecular mass fractions, which often impairs the solubility of a resin, can best be recognized by the Z-average molecular mass M_Z.

The resins have important impact on the aging resistance of the adhesives. Unsaturated sites are susceptible to peroxidation. Most resins reduce the efficiency of radical cross-linking considerably.

Usually the resins are delivered as solids in barrels, as broken pieces, or as pellets or flakes. Recently an increasing number are also available dispersed in water for water-based adhesives, and molten in tank trucks for hot-melt adhesives.

Rosin and Rosin Derivatives. The main component of rosins are rosin acids with two double bonds the molecule. They can be stabilized by hydrogenation or disproportionation. The softening temperature can be varied over a wide range by esterification, from the methyl ester (liquid) to the pentaerythritol ester (ca. 110 °C). Still higher temperatures are achieved by oligomerization. Rosin-based resins have a broad range of compatibilities.

Terpene resins are produced by cationic polymerization of terpenes from turpentine or other sources. Poly(β-pinene) resins have the best compatibility to natural rubber. They are fairly stable, in spite of the double bond in the cyclohexyl ring. Due to increasing demands for terpenes and decreasing availability, the resins are being substituted progressively by hydrocarbon resins.

Hydrocarbon resins are produced mainly by cationic polymerization ($AlCl_3$ and BF_3) of carefully selected fractions from mineral oil processing. There are two main types: aliphatic resins from C_5 and $(C_5)_2$ fractions, and aromatic resins from C_9 fractions.

The aliphatic resins have good compatibility with natural rubbers and polyisobutylene. The degree of unsaturation, which has an impact on aging and radical cross-linking efficiency, is characterized by the bromine number. Recently an increasing number of aromatically modified C_5 resins with improved adhesive properties in natural rubber and block copolymers are available. Some are hydrogenated to improve aging properties.

Aromatic C_9 resins have good compatibility with SBR and EVAc elastomers, and some of them to acrylate copolymers. They have limited compatibility with natural rubber. For more details, see [211, pp. 527–544].

2.4.3. Plasticizers

Plasticizers are generally liquid substances of low molecular mass (some 10^2 to a few 10^4) that lower the T_g of the adhesive. Thus, they often reduce peel adhesion at ambient temperature but increase low-temperature adhesion. They increase tack and creep of the adhesive.

Most important groups are:

- Mineral oils, which may be aliphatic, naphthenic or aromatic
- Oligomers with low T_g, such as polybutenes, liquid resins, polyethers, and branched polyesters
- Long-chain alkyl esters of aromatic and aliphatic acids such as phthalic, adipic, sebacic, and trimellitic acids, which are typical PVC plasticizers

2.4.4. Fillers

Fillers are usually powdered inorganic solids. Primary parameters are the average and distribution of particle size, as well as the particle shape. For very small particles, often the surface area per unit weight is given instead of this. The key factor is the degree of dispersion in the adhesive, which is determined by the surface energies, the mixing or kneading process, and additives.

Fillers increase creep resistance and improve removability of the adhesives. If they are used as dyes, colorimetric parameters are important. For special purposes, fibers, spheres, conductive particles, etc. may be added to PSAs.

The most important groups of fillers are:

- Carbonates, silicates, and silica are the most common fillers. Interaction with the elastomers can be modified greatly by surface treatment with fatty acids or silanes. Pyrogenic silica has an extremely high surface area.
- Zinc oxide promotes chemical cross-linking by phenolic resins and thio compounds.
- Glass and polymer fibers are used to decrease strain under the influence of unwinding stress in unsupported double sided tapes (transfer tapes).
- Glass and polymer microspheres improve the stress distribution in the adhesive layer and the adhesion of thick double-sided tapes [245]. Organic microspheres containing low-boiling liquids can be used to foam an adhesive layer [246].
- Organic and inorganic pigments are used as coloring agents. Most common is titanium dioxide for whitening. Surface treatment can prevent light induced decomposition of the elastomer.

2.4.5. Stabilizers

In terms of stability the polydiene elastomers are the most critical due to their main-chain double bonds. Oxidative degradation is stimulated by elevated temperatures and the presence of unsaturated resins, typically used as tackifiers.

Elastomers with saturated chains like polyacrylates and polyisobutylenes are considerably more stable.

Primary antioxidants act by trapping radicals and preventing the initiation of oxidative chain reactions. Sterically hindered phenols are widely valued for their low- or nonstaining properties. Secondary stabilizers (e.g., phosphite esters and special thioesters) decompose peroxides, which are responsible for autocatalytic effects.

Stabilization against UV can be achieved by UV absorbers such as derivatives of benzotriazoles and benzophenones, and by hindered amine light stabilizers (HALS). For more details, see [247].

2.4.6. Release Coatings

Release coatings are important components of pressure sensitive tapes as they enable the formulator to adjust the release force according to the application, for example, the unwinding force of tapes and the transfer of double-sided tapes to substrates.

The surface energy of a release coating is much lower than that of the adhesive. Thus the degree of wetting and adhesion are reduced. Further factors are the viscoelastic properties of the adhesive, the surface texture, and the number of pores in the release coat. External factors such as contact pressure, contact time and temperature should have as low an influence as possible.

There are three main types of release coatings:

- Release coatings based on long-chain alkyl groups yield medium release forces (0.1 – 1 N/cm). They are primarily used for the backsize of tapes. Examples are polyvinyl – stearyl – carbamate, chromium complex of C_{14} – C_{18} fatty acids [248], and stearyl copolymers [249].
- Acrylic polymers with perfluorated alkyl groups [250] are applied from solvent or as latex [251]. Their use seems to be rather restricted.
- The most common release coatings are silicones. They are based on poly(dimethylsiloxane), yield low to medium release forces (0.01 – 0.3 N/cm), and are applied primarily on release liners.

The key process for silicones is cross-linking, since residues of low molecular mass silicone oil on the surface of the PSA will eliminate tack.

The first reactions used were tin-catalyzed condensation reactions, which could be applied from solvents only. Since ca. 1980, noble metal catalyzed addition reactions have been introduced to coating technology. The combination with retarders gives systems which cross-link efficiently at high temperature (e.g., 10 s at 150 °C) but are

Table 3. Thickness and viscosity of adhesive tape coating layers

	Thickness, μm	Viscosity, ** Pa · s
PSA	10 to 150 *	0.5 – 100
Primer	0.5 to 5	0.01 – 1
Release layer, backsize	0.1 to 2	0.001 to 5

* For special double-sided tapes, up to 2 mm. ** Upper values for 100 % systems.

fairly stable at ambient temperature. They can be applied without solvents. For more details, see [211, p. 601 – 626].

The influence of high temperatures can be avoided by radiation curing (UV and EB). There are two different processes: one is initiated by radicals [227, p. 355 ff] and the other one by ionic species [254]. Both processes allow the release force to be adjusted by means of additives to achieve controlled release.

2.5. Manufacturing Processes

Adhesive tapes generally consist of flexible backings with layers of primer, adhesive, and release coatings. Thus, manufacturing adhesive tapes essentially comprises compounding, coating and slitting processes.

The thickness of the layers and the viscosity of the coating materials vary over a broad range. Typical values are listed in Table 3.

The tolerance in thickness often has to be kept below ± 10 %. Typical line speeds are in the range of 50 to several 100 m/min.

In the beginning of the PSA industry, most compounding and coating was effected in solution. This still is the most versatile process. With increasing demands for environmental protection, solvent recovery and incineration have become widespread. They are rather effective, but require high investment and increased energy costs. Therefore, they are being substituted increasingly by solventless processes.

Coating of fluids such as primers and release agents are well described in the literature. A guide to selection is given by Cohen [255, p. 13 ff]. Important parameters are the required thickness and precision of the layer as well as the desired line speed, the properties of the coating (e.g., viscosity, solvent content, shear tolerance) and of the substrate (e.g., tolerance of thickness, roughness, absorbency).

Pressure-sensitive adhesives are compounded and coated in the fluid state and used in the solid state. Key characteristics of the process are the fluidization and solidification procedures. Four alternative types of process can be applied (Table 4).

These divergent processes require very different types of equipment. The most important are summarized in Table 5. A comprehensive treatment is given in [211, pp. 355 ff].

Table 4. Alternative manufacturing processes for PSA coatings

Process type	Fluidization by	Solidification by	Strong points	Critical points
Solvent-based	solvation	evaporation	most versatile adhesive properties full range of thickness	high cost for energy and equipment residual solvent, especially at higher thickness influence on some backings
Waterborne dispersions	dispersion in water	evaporation and coagulation	thin to medium thickness high cohesion applicable to solvent-sensitive films	low tack on plastics and rough substrates low water resistance electrochemical corrosion
Hot melt	heating	cooling	medium to very high thickness high peel adhesion high cohesion at ambient temperature high speed	low cohesion at higher temperature, (can be overcome by cross-linking) thermal and oxidative degradation
Reactive systems	(liquid components)	curing (chemical or radiation initiated reactions)	thin to very high thickness high cohesion good reversibility	low tack on rough substrates

Table 5. Alternative equipment for different manufacturing processes

Solvent-based	Waterborne	Hot melt	Reactive system
Compounding			
granulator + kneader (solid rubbers) mixer (soluble elastomers)	mixer	kneader (discontinuous) mixer extruder (continuous)	mixer
Coater			
knife over roll reverse roll slot die	Meyer bar air brush reverse roll slot die	slot die melt roll extrusion roll calender	reverse roll slot die
Drying			
dryers + solvent recovery or incineration	dryer (air cleaning often necessary)	none	none
Cross-linking/curing			
high temperature radiation	high temperature radiation	radiation	mixing two components radiation

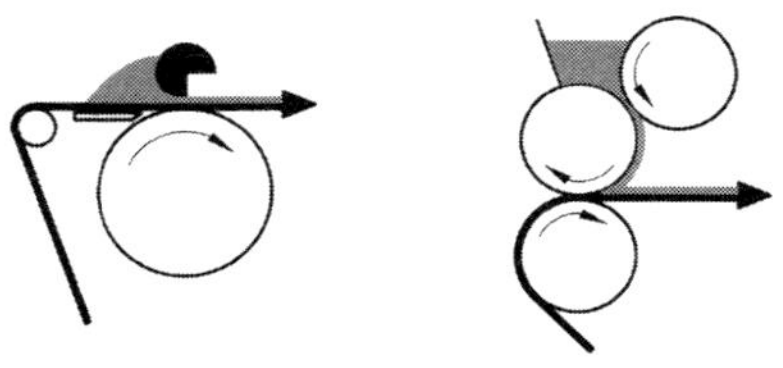

Figure 15. Standard coating equipment for PSAs
A) Knife over roll; B) Reverse roll coater

2.5.1. Compounding

Bales of natural rubber can be processed in two ways:

1) With low degradation: Granulating the bales and kneading them together with all other ingredients in solution kneaders with sigma blades. If high-molecular rubber types like SMR-L (Standard Malaysian Rubber, light), ADR (Air Dried Rubber), or crepe are used, the resulting PSAs can be applied for general purpose tapes without cross-linking. Primary disadvantages are slow batch processing and low solids content or high viscosity of the coating materials.
2) With high degradation: The rubber bales are milled down together with the filler in an internal mixer with intermeshing blades within a few minutes. Subsequently this mill base is dissolved and compounded with tackifiers and further ingredients in a solution mixer. Process time is much shorter, but cross-linking is indispensable even for tapes with standard performance.

Compounding block copolymer based hot-melt adhesives can be performed batch-wise in heated kneaders or continuously in mixing extruders [256], [257]. A survey of the designs and parameters of suitable extruders is given in [258].

Compounding of soluble elastomers, dispersions, and liquid components is described in detail in [255, pp. 23–61].

2.5.2. Coating

Standard equipment for solution based PSAs are knife over roll and reverse roll coaters (Fig. 15) [255, p. 13 ff]. For high-speed application, a closed reservoir or a slot die is preferred for better control of air bubbles [260]. Especially the knife over roll coater allows for an easy change of the coating material.

The coating can be applied directly to the web or to a release liner and subsequently transferred to the web. The latter method is often used for double-sided tapes with foam backings, for it eliminates the influence of solvent and high temperature on the substrate.

For hot melts, roll coaters and die systems are used (Fig. 16). Typical coating temperatures are in the range of 120 – 160 °C.

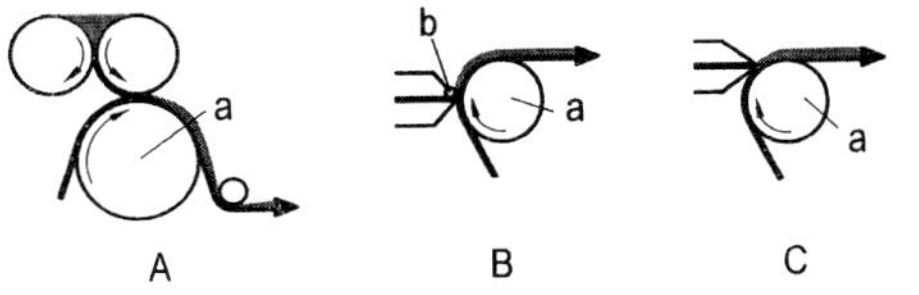

Figure 16. Coating equipment for pressure-sensitive hot-melt adhesives
A) Roll coater; B) Slot die coater; C) Extrusion die coater; a) Chill roll; b) Rotating rod

Roll coaters are largely free from interference caused by particles and variations in web thickness. The coating material can be changed quite easily. However, the influence of oxygen on the adhesive cannot be excluded completely, resulting in higher degradation of polydiene block copolymer hot-melt adhesives.

There are two types of coating dies: In the *slot die* coaters the coating is pressed onto the substrate. The thickness of the layer is controlled by the gap between the die lips and the web, which is backed by a steel or hard rubber roller. At the forward lip there may be a rotating rod to reduce stripes caused by particles. The pressure improves the contact of the adhesive to the substrate. The thickness of the coating is influenced by the web thickness. Further details are given in [261].

In the *extrusion die* coater, the coating layer is formed and calibrated in the slot and laid down on the substrate. The thickness is independent of the web thickness and can be reduced by stretching. There is no penetration through open substrates like fabrics, and there is no influence of particles on the coating. Good contact to the substrates can be provided by an air knife or an electrostatic field.

2.5.3. Drying

In solvent- and water-based systems the drying process is often the limiting factor for the line speed. Thus, it determines the economical efficiency of the production line and has to be considered carefully for all products.

Up to coating weights of about 50 g/m^2, forced air drying tunnels (Fig. 17 A) are employed. The air is pressed through holes, slots, or nozzles onto the web. Usually the dryers are segmented to cope with the different drying regimes [255, pp. 267–298]. In single-side dryers, the web is backed by rolls or by a belt. Most efficient are two-sided floater dryers, in which the web hovers between specially designed air nozzles on both sides.

For higher coating weights, festoon dryers (Fig. 17 B) can be used. In this type of dryer, long loops of the web hanging on rods are moved slowly through a sealed room, which may have a volume of more than 500 m^3. The air is pressed through nozzles from the ceiling into the loops. Further details are given in [255, pp. 86–91].

In all air dryers, the concentration of the solvents must be kept below 50 % of the lower explosion limit (e.g., ca. 20 g/m^3 for hexane and toluene, ca. 40 g/m^3 for ethyl acetate [262]). In high-performance lines huge amounts of air have to be heated, transported and finally purified of solvent, for example, by adsorption processes.

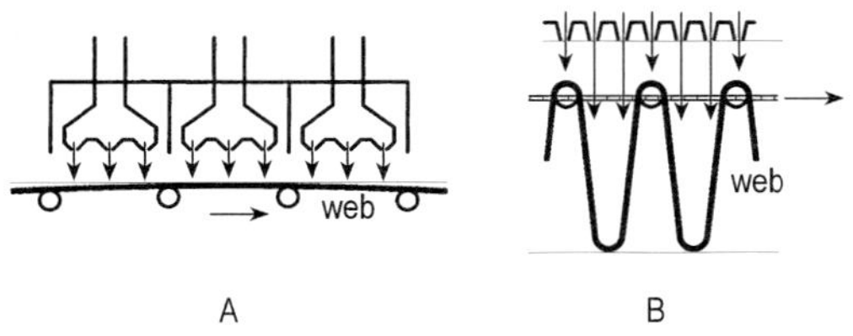

Figure 17. Drying equipment for pressure-sensitive adhesive tapes
A) Drying tunnel; B) Festoon dryer

Table 6. Characteristics of EB and UV radiation

	EB radiation	UV radiation
Particle/quantum energy	150 – 300 keV	3 – 6 eV
Generation of radicals	stochastically by all matter	by the photoinitiator (PI) only
Absorption of the radiation	by all matter	by specific components
Penetration depth	dependent on density and particle energy	dependent on wavelength and concentration of components, especially the PI

An alternative method is incineration. As the solvent concentration is low, effective heat recovery equipment is necessary to reduce the amount of additional fuel [255, pp. 110 – 115]. Catalytic oxidation often is inhibited by widely used substances like silicones and amines.

Energy consumption can be reduced if oxygen is excluded from the dryers. This can be achieved by substituting the air by nitrogen in a closed loop. In this case the solvent concentration can be greatly increased, so that the flowing gas volumes are reduced considerably and the solvent can be condensed at about – 5 °C. The disadvantages of this method are the limited accessibility of the coaters and drying tunnels and the need to seal the inlet and outlet of the web sufficiently.

2.5.4. Cross-linking / Curing

Cross-linking can be effected by chemical reactions of reactive groups in the elastomer among one another or with reactive additives. The reaction must not start until after the coating process. This is effected by high temperature, which may be achieved in the last sections of the drying tunnel or in a separate high-temperature tunnel.

Radiation curing is very appropriate to cure PSAs, as the transferred energy is far higher than in thermal cross-linking. Both electron beam (EB) and ultraviolet (UV) radiation are applied. Their physical characteristics differ, as shown in Table 6 [263]:

These differences result in different kinetics. The energy of EB radiation is several tens of thousands times higher than the energy of chemical bonds. A primary electron transfers its energy in a cascade of secondary and tertiary electrons to several thousand molecules, generating reactive sites stochastically in the elastomers, the resins, and all other additives. The radical sites on the elastomers result in cross-linking of the adhesive. In the presence of reactive monomers, there is a high degree of grafting.

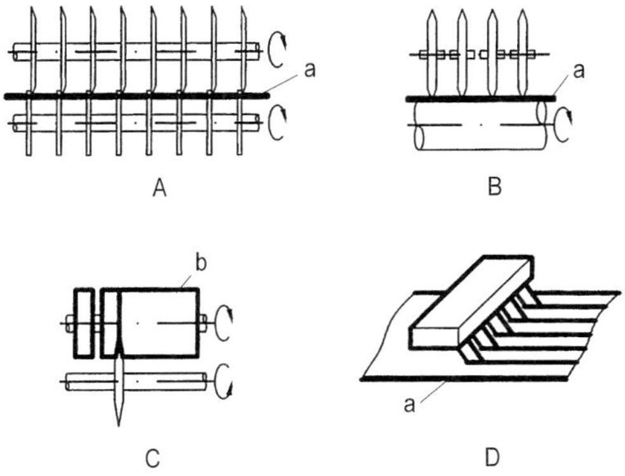

Figure 18. Slitting methods for pressure-sensitive adhesive tapes
A) Shear cutting; B) Crush cutting; C) Log cutting; D) Razor blade cutting
a) Web; b) Log

The quantum energy of UV radiation is comparable to chemical bond energies. The radiation is absorbed by the photoinitiator, which is converted to free radicals or to a highly reactive state. These sites generate cross-links at the elastomers; in the presence of monomers, they can initiate polymer chains.

The other difference concerns penetration depth. As EB radiation is attenuated by all matter, penetration depth is a function of electron energy and density, e.g., at 300 kV the useful penetration depth at a density of 1 g/cm^3 is 0.35 mm [264]. In a clear layer of nonabsorbent polymers the penetration depth of UV radiation may increase with the absorbed irradiation energy (i.e., irradiation time) [265]. With low concentration of the photoinitiation, optimal wavelength, and sufficient irradiation time, coatings with a thickness of more than 1 mm can be cured.

Details about equipment and process parameters are given in [266], [267].

2.5.5. Slitting, Cutting

For high efficiency, coating usually is performed over a width of more than a meter. The tape rolls are obtained by cutting the web into strips with a width of a few millimeters up to several centimeters. Four methods are applied (Fig. 18).

Shear cutting (Fig. 18 A) is effected between a rotating knife and a roller edge and results in tapes with smoothly cut edges. This method has the highest precision, but the width of the tape rolls cannot readily be changed. It is used for most rigid-film tapes and double-sided tapes.

Crush cutting (Fig. 18 B) is effected by pressing a rotating knife against a steel roller. The edges of the tapes are roughened by small cracks. The precision is lower than that of shear cutting, but the width can be changed more easily. This method is usually applied for paper-, fabric-, and nonwowen backed tapes. In a special case biaxially oriented polypropylene (BOPP) based films are cut by such a process to achieve hand-tearable tapes.

In log cutting (Fig. 18 C) the coated web is wound into rolls or logs in full width and in the desired length of the tape rolls. Subsequently, these logs are sliced by a rotating knife, which is moved transversally between slices to match the width of the tape rolls. As the transverse location can be adjusted deliberately by a computer, this process allows the complete customization of the width of the tape rolls.

In razor blade cutting (Fig. 18 D) the web is cut by parallel razor blades. The edges are cut very smoothly. Due to the lower stiffness of the blades the precision is lower than with the other methods. It is preferred for, e.g., carton-sealing tapes.

2.6. Adhesive Tape Products

Hundreds of divergent pressure-sensitive adhesive products are available to cope with numerous applications in industry, office, household, etc. They differ in material and/or in design. Most are wound into rolls; some are die-cut into separate pieces or assembled on sheets.

A detailed presentation of all the products and applications is beyond the scope of this article. Reference is made to the *Handbook of Pressure Sensitive Technology* [211].

2.6.1. One-Sided Adhesive Tapes

Generally such tapes are composed of four layers:

1) Pressure-sensitive adhesive
2) Primer
3) Backing (carrier)
4) Backsize (release coating)

The backsize and the primer are optional components. The former is used to adjust the unwinding characteristics and/or to seal a porous backing material. The latter provides for good bonding between adhesive and backing. In many cases, instead of using a primer the film surface is treated by corona discharge or with a flame. Sometimes the primer also contains an agent which cross-links the adhesive by diffusion into this layer.

2.6.1.1. Film Tapes

Film tapes have by far the largest production volume. According to the deformation properties there are two kinds of film tapes.

Rigid Film Tapes. *Cellophane* is the oldest transparent tape backing. It is fairly strong and transparent and is widely used to fasten package wrappings, to seal cartons, and for a great variety of other sealing, holding, and mending jobs. However it is not resistant to moisture and has become rather expensive due to environmental protection regulations in the production process.

Cellulose acetate film is resistant to moisture but lacks the toughness of cellophane. In combination with an age-resistant acrylic adhesive and a matte backside it is used in many countries for office tapes, to repair torn books, etc.

Considerable amounts of oriented *poly(vinyl chloride)* (PVC) films are used for tapes in Europe. In combination with solvent-based rubber – resin adhesives, they yield high-performance packaging tapes with smooth unwind characteristic without the need for a backsize.

Biaxially oriented polypropylene (BOPP) tapes with hot-melt adhesives based on styrenic block copolymers have now attained the biggest share of the market for carton-sealing tapes. They require a backsize, and on unwinding they evolve much noise.

By using waterborne acrylic dispersions, UV-resistant packaging tapes with quiet unwinding are produced. Most of them are used as carton-sealing tapes. Further applications are protection of labels and documents.

By stretching PP films predominantly in the machine direction, the tensile strength in this direction can be increased further at the expense of strength in transverse direction. Such tapes are referred to as machine direction oriented polypropylene (MOPP) and tensilized polypropylene (TPP) films and are used for strapping and bundling heavy goods, to secure pallets, and to tear the corrugated cardboard to open boxes. These tapes often are wound into special rolls with lengths of up to 70 000 m by cross-spool winding.

For more information on packaging tapes, see [211], [268].

The upper temperature limit of a film tape is given by the softening of its backing, which may be caused by the glass transition (e.g., PS, PVC films) or by the melting of crystalline regions (e.g., PE, PP, PETP films). The low temperature limit is given by embrittlement of the film, or more often by the glass transition of the adhesive, which is not considered here.

To assess roughly the temperature range of the most important film backings some values are given in Table 7, valid for adhering films at low stress. It should be noted, that for long-term service, the upper limit may be reduced considerably due to degradation processes and interaction with the adhesive. This is treated in more detail in Section 2.6.1.6.

A comprehensive survey on further film properties is given in [269].

Producers and Trade Names: Asia Chemical Corp., Beiersdorf (tesa tapes), Intertape Polymer Group, 3M (Scotch tapes), Scapa Tapes Klebetechnik (Cellux, Sellotape), Vibac.

Conformable Film Tapes. For many applications the tape should conform to a rugged or profiled surface. There are two types of films, which differ in their tensile stress elongation curve.

Table 7. Application temperature range (short-term) of important film backings [269], [270]

Film	Application range, [a], °C	
	min.	max.
Cellophane	0	150
Low-density Polyethylene [b]	−50	95
High-density Polyethylene [b]	−40	80
Oriented Polypropylene	0	100
Rigid PVC	−15	75
Plasticized PVC [b]	−30	100
Poly(ethylene terephthalate) [c]	< −100	160
Poly(ethylene naphthenate) [c]		200
Poly(ether ketone)		240
Polyimide	< −100	350

[a] Temperature range over "which no significant loss in strength or stiffness occurs" [269]. [b] Conformable films. [c] Biaxially oriented.

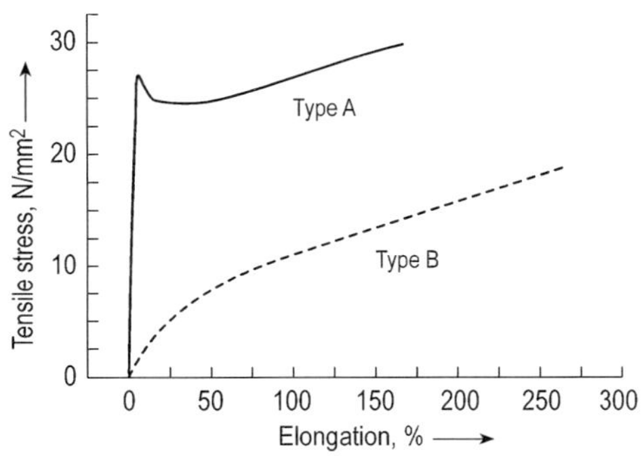

Figure 19. Deformation characteristics of conformable films

Type A tapes are made of semicrystalline polymers like PE. On extending such polymers beyond the yield point, the film deforms irreversibly at nearly constant stress until it finally increases to the breaking point. Type B tapes are made of amorphous plastics like plasticized PVC or rubbery EPM copolymers. Here the stress increases steadily up to the breaking point. It is a rubberlike reversible deformation. There is always a residual stress in the adhesive bond. Therefore the adhesive needs sufficient cohesion. Typical values are shown in Fig. 19.

Tapes of type B are used for insulating and bundling wires and for cable harnessing. Plasticized PVC usually contains low molecular mass plasticizers, which may migrate into the substrates and adhesive. This can cause swelling of the substrate, and the tape may become brittle due to loss of the plasticizer.

Conformable films of type A are used as film-masking tapes and as thick films for corrosion protection tapes for underground transmission pipes.

Producers and Trade Names: Beiersdorf (tesa tapes), Coroplast Fritz Müller KG, 3M (Scotch tapes), Nitto Denko Corp., Raychem, The Tapecoat Company.

2.6.1.2. Fiber-Reinforced Tapes

To increase the mechanical strength of a film tape further, filaments can be embedded into the PSA. Most common are glass fibers; alternatives are polyester and polyamides. Such tapes have high tensile strength and low elongation under stress. Further, they are highly resistant to tear propagation, which is a weak point of biaxially oriented films. These tapes are suitable for heavy-duty applications.

Producers and Trade Names: American Tape Company, 3M (Scotch tapes).

2.6.1.3. Paper Tapes

Paper is a widely used backing for pressure-sensitive adhesive tapes. It is more expensive than simple plastic films but has several advantages. It is easy to tear by hand, tear strength is sufficient for many applications, it does not soften at high temperatures, and it can be upgraded by coating or saturating with polymers.

Flat Kraft Paper Backing. Usually the back side of the kraft paper is coated with PE by extrusion and additionally with a silicone release layer. The other side is coated with the PSA. Tapes with regular kraft paper have low elongation at break (ca. 4 %). This can be increased to ca. 10 % by corrugating the paper using the Clupak technology. Kraft paper tapes are quite common in Japan for carton sealing and general purpose packaging.

Creped paper backings can be divided roughly into two groups.

Thick creped papers with a thickness of more than 300 μm can be used without a backsize coat for high-elongation masking tapes for painting.

Flat creped papers saturated with rubberlike polymer latices generally have a backsize coating to control the unwind characteristic without impairing the adhesion of paints. In combination with natural-rubber-based adhesives, these tapes are used for masking in the automotive industry, for assembling electronic components, and insulation of low-voltage devices etc.

For high-temperature paint curing, the adhesive is cross-linked and thermosetting, i.e., it contains reactive resins, which produce additional cross-links during application in the high-temperature drying tunnels.

Producers and Trade Names: American Tape Company, Beiersdorf (tesa tapes), 3M (Scotch tapes), Nitto Denko Corp., Permacel, Shuford Mills, Sicad.

2.6.1.4. Textile Tapes

Fabric-based PSA tapes combine low stiffness with high tensile strength and low elongation. Usually they are tearable by hand. The backside can be open, coated with an acrylic finish, or extrusion coated with PE copolymers. *Duct tapes* are produced by combining a rubber-based adhesive, a wide-mesh fabric, and a PE film by calendering in a one-step process. Fabric tapes are used for a vast number of divergent applications.

For many applications *nonwovens* or *fleeces* are used. Such tapes may be coated with a porous or impervious backsize.

Producers and Trade Names: Beiersdorf (tesa tapes), 3M (Scotch tapes), Nichiban, Scapa Tapes Klebetechnik (Fermoflex, Rotunda).

2.6.1.5. Other Backings

Foams. Very common are polyurethane foams. Foams based on polyesters have better heat and light stability than polyethers but have inferior stability against moisture. To increase light stability, they may contain carbon black. The PU foams are increasingly being substituted by foams based on PE and PE copolymers. For very high temperatures, silicone-based foams are available. The main applications of foam tapes are sealing gaps and damping vibrations.

Producers and Trade Names: Illbruck Industrieprodukte, Norton Company, Rubber Astic, Venture Tape Corporation.

Metal Foils. Most common is aluminum, less common are copper and lead. Such tapes are quite conformable and are used to seal against moisture and water. Tapes with aluminum foils or aluminum-coated films can shield surfaces against thermal radiation. Copper and aluminum foils and vapor-coated films also can be used to shield against electromagnetic interference.

Producers and Trade Names: Coroplast Fritz Müller, 3M (Scotch tapes).

2.6.1.6. Electrical Tapes

Electrical tapes comprise different types of backings. Conveniently they are treated in a separate section as they have to fulfill common specific requirements e.g. good electrical properties under dry and moist conditions, long service life at elevated temperatures, and often they need flame resistance and must be free from any substance corrosive to metal surfaces and from any emission harmful to electrical contacts.

These requirements are described by national and international standards like PSTC [220], UL [271], ASTM [221], VDE [272], and DIN [273]. The International Electrotechnical Commission (IEC) [274] has established a worldwide network to unify all national standards and test methods concerning electrical tapes.

As temperature often is the dominant aging factor in electrical insulating systems, certain basic thermal classes have been recognized internationally (Table 8).

Within each class certain combinations of backings and adhesives are feasible. For long-term service above 150 °C, silicone-based PSAs must be used.

For more details see Bräunling [211, p. 643 ff] and Külper [275].

Producers and Trade Names: 3M (Scotch tapes), Nitto Denko Corporation, PPI.

Table 8. Thermal classes for electrical tapes for long-term service life at elevated temperature

Thermal class	Temperature, °C	Typical backings
Y	90	stabilized PVC
A	105	saturated paper, cotton fabric, acetate film
E	120	
B	130	poly(ethylene terephthalate) film, fabric, nonwoven
F	155	poly(ethylene naphthenate)
H	180	polyimide, glass fabric

2.6.2. Double-Sided Adhesive Tapes

Double-sided tapes usually are used to attach substrates to one another; therefore, the adhesive is of primary importance. Generally, the backing material, if present, must assist the handling.

In high-performance tapes the backing consists of a relatively thick highly compliant cross-linked polymer layer, which may be foamed, and has a major impact on the strength of the bond.

Generally a double-sided PSA tape comprises five layers together with a release liner consisting of three additional layers:

- Pressure-sensitive adhesive 1
- Primer 1
- Backing (carrier)
- Primer 2
- Pressure-sensitive adhesive 2
- Release coating 1
- Release liner
- Release coating 2

Primers are optional, and in case of polyolefins can be substituted by corona, flame, or fluorine treatment.

2.6.2.1. Standard Double-Sided Tapes

Most double-sided tapes have cross-linked acrylic PSAs, which often are tackified with compatible resins. The adhesives on both sides can be identical or different, for example, to combine reversible and permanent bonding. Some adhesives are based on natural rubber or styrenic block copolymers.

Widely used backings are rigid and conformable films, nonwovens, fabrics, creped paper, and foams based on polyurethanes, polyolefins and their copolymers, etc. Generally the thickness of the adhesive layer has to be increased with increasing roughness of the substrates. Foams are especially useful for rugged or profiled surfaces.

An important element is the interliner, which mainly is coated with a silicone release agent, as described in Chapter 2.4. Glassine kraft paper is used most frequently. Creped paper, rigid and semirigid films are also used. Besides the peel force, stiffness and conformability are significant features of the interliners. More detailed information is in [211, p. 691 – 705].

Double-sided tapes are used for a vast diversity of applications in industry and household. Further important applications are laminating narrow profiles and wide substrates like foams and thick mats to endow them with pressure-sensitive adhesion. Two special applications are:

1) In the papermaking and printing industry splicing tapes are used to bind the end of one roll of paper to the beginning of another. This must be effected at high speed, and the bond often has to bear high temperature. Moreover, as the tape has to be returned to the papermill for reprocessing, the adhesive and the backing have to be repulpable (i.e., soluble or dispersible in water). Such PSAs are based on highly acidic polyacrylates and compatible water-soluble plasticizers. The dispersibility is adjusted by partial neutralization with alkali [276].
2) Double-sided tapes are used further to fix the printplates onto the printing rollers. The adhesive has to ensure that the tape is mounted easily and the plate stays down during the printing process. After finishing the plate must be easily demountable without residues.
   Very important for the print quality is the type and hardness of the backing: Film backings are mainly used to print solid areas. Foam backings are used for half-tone and combination prints. The backing has to exhibit a high recovery rate to achieve high printing speed with constant quality.

Producers and Trade Names: Beiersdorf (tesa tapes), Lohmann (Duplocoll tapes), MACtac, 3M (Scotch tapes), Nitto Denko Corporation.

2.6.2.2. Transfer Tapes

In transfer tapes, the PSA is coated on the release liner without additional backing. Therefore the release forces on both sides of the liner have to be precisely different and constant over storage time. The adhesives sometimes contain short-cut glass fibers to decrease the elongation during unwinding. Such tapes are used to transfer the PSA to extruded profiles or other parts to endow them with pressure-sensitive adhesion. For use in household and office, special dispensers are available.

Producers and Trade Names: Avery-Dennison Corporation (Fas tapes), Beiersdorf (tesa tapes), MACtac, 3M (Scotch tapes), Scapa Tapes Klebetechnik (Cellux).

2.6.2.3. Adhesive Pads and Strips

Adhesive pads and strips are die-cut pieces of relatively thick layers of highly conformable but cohesive PSA covered with release liners on both sides. Often they have a backing or core of polymeric foam. They are used to fasten objects to many surfaces.

Adhesive strips are constructed so, that in spite of their high adhesion, they can be removed easily from all surfaces just by stretching them in the bond direction. This principle first was used for medical applications [277]. More recent developments allow objects weighing up to a few kilograms reversibly to be fastened to various surfaces [278].

Producers and Trade Names: Beiersdorf (tesa Power Strips), 3M (Scotch Command Tack).

2.6.2.4. High-Performance Tapes

Peel strength and adhesion under bending load can be increased by combining a highly cohesive PSA with a tough conformably balanced viscoelastic backing of adapted thickness. Such tapes also give very high bond strength on rough or profiled surfaces. Adhesion can be promoted further by application of pressure and elevated temperature (30 – 50 °C) during the bonding process and by careful exclusion of air bubbles at the interface.

The first products were based on UV-cured acrylics with a thickness of 0.3 to a few millimeters filled with glass microbubbles [279]. This has been extended now to a range of products for various interior and exterior applications, e.g., in the construction, automotive, building, and electronic industries. Such performance can be achieved also by polyurethanes [280] and other elastomers.

Quite recently the performance of PSAs has been upgraded further by introducing thermosetting acrylic adhesives. This is achieved by additional epoxy groups in the PSA, which are initiated by heating after application of the tape [281].

Producers and Trade Names: Lohmann (Duplocoll tapes), 3M (Scotch VHB tapes), Norton Company.

2.6.3. Medical Tapes

Medical tapes are the oldest application of pressure-sensitive tapes [203]. Major progress towards skin compatibility was made by introduction of zinc oxide into the rubber-based PSAs [282]. With the development of polyisoprene, polyisobutylene, and other synthetic elastomers, natural rubber was replaced to a certain extent in medical products by synthetic rubbers. These were followed by the styrenic block copolymers [211, pp. 317 – 373].

In the late 1950s, polyacrylate-based PSAs were developed [283] and gained a dominant market share in the following decades. To a smaller extent poly(vinyl ethers) are used in medical devices. For specific applications like transdermal drug delivery systems, ostomy mounts, and electrically conductive adhesives a variety of specialty polymers such as silicones, polyurethanes, and various hydrogels are applied.

Nevertheless, both rubber- and acrylate-based PSAs are still important for medical products. Rubber-based adhesives have high tack in combination with a medium cohesive strength. The advantage of acrylics is their low allergenicity and the possibility to polymerize tailor-made products for particular applications. This is achieved by the choice of type and ratio of monomers for the polymerization and cross-linking process.

Basic characteristics of PSAs for medical applications are: Fast wetting of the skin during initial adhesion, secure adhesion on various skin types during application, and complete and pain-free removability. For resistance to moisture such as water or sweat, good permeability to water vapor and air is required.

Pressure-sensitive adhesive products are utilized in a broad variety of medical applications in very different fields: First aid products for consumers, medical tapes, island dressings, surgical drapes, wound-closure strips in hospitals and physician's practices, as well as tapes and bandages for orthopedic and athletic use. Other uses include transdermal systems for therapeutic purposes, and ostomy mounts, attaching electrocardiogram devices, and protective foot products.

First-Aid Bandages. The most widespread medical PSA products are the small bandage strips. This bandage is composed of a backing material coated with an adhesive and a small absorbent wound pad [284]. The adhesive layer is covered by a release paper or plastic film which is removed before use. For this kind of products plastic films (e.g., polyolefin), rigid or elastic fabrics, and nonwovens are applied. To attain an appropriate permeability, in many cases the backing material is perforated.

Hospital Tapes. Many hospital products are used under sterile conditions. Therefore, the raw materials have to be resistant to different sterilization methods. For several medical devices, transparency to X rays enables examination without removing the bandage.

Depending upon the nature of the backing material, it is possible to manufacture occlusive as well as vapor-permeable products. Permeability to water vapor results in a better skin adhesion because the accumulation of moisture between skin and adhesive is avoided [285].

Medical tapes are used for all kinds of fixing purposes. The carrier of medical tapes can be woven fabric, film, perforated film, laminates, and nonwovens.

Island dressings applied after surgery are designed to absorb wound fluid either in their wound pad or in especially designed adhesive systems like hydrogels and hydrocolloids. Others are made of translucent films to enable inspection of the wound without removal of the dressing.

Self-adhering surgical drapes assure a sterile environment for surgery by isolating the operative area from resident skin bacteria. Due to the very elastic and flexible properties of the films (e.g., polyolefin or polyurethane) and their thin adhesive layer, they conform tightly to even difficult anatomical contours.

Wound-closure strips cover the field of small incisions or injuries which easily can be closed without stitches. The strips are made from a rigid fabric coated with a highly biocompatible PSA.

Orthopedic Tapes. In orthopedic applications, PSA products are most frequently used as tapes and elastic adhesive bandages. The tapes consist of a rigid backing material (e.g., cotton fabric) coated with the PSA. They are used in therapeutic medical applications to control motions and to immobilize and to stabilize joints in athletic use. Orthopedic bandages are made from coated elastic material and often are used when compression and support are to be supplied to the extremities.

Transdermal Systems. The use of transdermal drug delivery systems in pharmaceuticals has also become an important area for PSAs. This noninvasive form of drug application allows for a controlled release of bioactive substances. On applying the patch to the skin, the release of the active ingredient starts immediately and stops upon removal. During application the PSA has not only to ensure the proper adhesion of the product but also the drug delivery into the skin. This may either be achieved from a reservoir through the adhesive (membrane system) or from an adhesive that itself contains the drug (matrix system). In transdermal drug delivery systems [286] highly biocompatible adhesives with appropriate compatibility to the drug are used. These are mainly polyacrylates, synthetic rubbers, and silicone adhesives.

Electrically conductive adhesives are used in medical applications such as electrocardiogram electrodes and grounding systems for electrosurgery [211, p. 637]. They consist of electrically conductive gels, i.e., moist hydrophilic polymers containing dissolved salt.

Producers and Trade Names: Beiersdorf, Bristol-Myers Squipp, Johnson & Johnson, Lohmann, 3M, Paul Hartmann, SCA Mölnlycke, Smith & Nephew.

2.7. References

[1] *Neue polymere Werkstoffe 1969-1974,* Hanser Verlag, München - Wien 1975.

[2] M. Michel: *Adhäsion und Klebtechnik,* Hanser Verlag, München 1969.

[3] National Materials Advisory Board, National Research Council: Treatise of Adhesion und Adhesives ("Structrual Adhesives with Emphasis on Aerospace Applications"), vol. 4, Marcel Dekker, New York-Basel 1976.

[4] J. J. Bikerman: *The Science of Adhesive Joints,* 2nd ed., Academic Press, New York-London 1968.

[5] C. Heitler in D. J. Alner (ed.): *Aspects of Adhesion,* **vol. 4,** University of London Press, London-Liverpool-Prescot 1968, p. 73.

[6] K. W. Allen in D. J. Alner (ed.): *Aspects of Adhesion,* **vol. 5,** University of London Press, London-Liverpool-Prescot 1969, p. 11.

[7] N. J. DeLollis in Ch. V. Cagle (ed.): *Handbook of Adhesive Bonding,* McGraw-Hill, New York 1973, pp. 2/1 – 2/16.

[8] W. C. Wake, *Polymer* **19** (1978) no. 3, 291.

[9] B. V. Deryagin, N. A. Krotova, V. P. Smilga: *Adhesion of Solids,* Consultants Bureau, New York-London 1978.

[10] W. A. Zisman, *Ind. Eng. Chem.* **55** (1963) 18.

[11] L. H. Shorpe, H. Schonhorn, *Adv. Chem. Ser.* **43** (1964) 189.

[12] R. J. Good in R. L. Patrick (ed.): *Treatise on Adhesion and Adhesives,* **vol. 1,** Marcel Dekker, New York 1967, pp. 9 – 65.

[13] D. H. Kaelble: *Physical Chemistry of Adhesion,* Wiley-Interscience, New York 1971.

[14] R. H. Ottewill in D. J. Alner (ed.): *Aspects of Adhesion,* **vol. 2,** University of London Press, Liverpool-London-Prescot 1966, p. 9.

[15] W. A. Zisman in P. Weiss (ed.): *Adhesion and Cohesion,* Elsevier, Amsterdam-London-New York 1962, p. 176.

[16] R. Köhler: "Fortschritte der Klebtechnik in den letzten zehn Jahren," *Chem Ing. Tech.* **42** (1970) no. 9/10, 599.

[17] A. Korolew, V. J. Bek, N. A. Grishin, *Vysokomol. Soedin.* **4** (1962).

[18] A. H. Walter, *Plaste Kautsch.* **17** (1970) no. 8, 601.

[19] S. S. Voyutskii: *Autohesion and Adhesion of High Polymers,* Interscience, New York-London-Sydney 1963.

[20] R. M. Vasenin: *Adhesion, Fundamentals and Practice,* Ministry of Technology, MacLaren, London 1969.

[21] J. R. Huntsberger in R. L. Patrick (ed.): *Treatise on Adhesion and Adhesives,* **vol. 1,** Marcel Dekker, New York 1967, p. 136 – 137.

[22] R. Köhler, *Adhäsion* **3** (1972) 66.

[23] W. Brockmann: "Grundlagen und Stand der Metallklebtechnik," *VDI-Taschenbücher T 22,* VDI-Verlag, Düsseldorf 1971, pp. 6 – 9.

[24] DIN 16920, Klebstoffverarbeitung (Processing of Adhesives), June 1981.

[25] S. Semerdjiev: *Metal-to-Metal Adhesive Bonding,* Business Books, London 1970.

[26] J. Shields: *Adhesives Handbook,* 2nd ed., Newnes-Butterworths, London 1976.

[27] Inter. Isocyanate Institute Inc.: *Recommendations for Handling Aromatic Isocyanates,* New York 1976.

[28] *Adhes. Age* **25** (Aug. 1982) 30 – 31.

[29] I. Skeist: *Handbook of Adhesive* 3rd ed., Van Nostrand Reinhold, New York 1985.

[30] R. Köhler, *Kunststoffe* **48** (1958) 441.

[31] R. Köhler, *Adhäsion* **16** (1972) 41.

[32] H. Lucke: *Kunststoffe und ihre Verklebung,* Verlag Brunke Garrels, Hamburg 1967.

[33] D. L. Bateman: *Hot Melt Adhesives,* 3rd ed., Noyes Data Corp., New Jersey, USA, 1978.

[34] Teroson, DE 24 54 235, 1974 (Kösters).

[35] G. F. Abele: *Hochfrequenz Schweißtechnik,* Zechner-Hurlig Verlag, Speyer 1973, pp. 387 – 391.

[36] D. Sates (ed.): *Handbook of Pressure-Sensitive Adhesive Technology,* Van Nostrand Reinhold, New York 1982.

[37] J. A. Radley (ed.): *Industrial Uses of Starch and its Derivates,* Applied Science Publ., London 1976.

[38] Blattman & Co., CH 256 502, 1949 (Meier).

[39] Henkel, DE 1 811 466, 1978 (Muszik, Dierichs).

[40] R. Houwink, G. Salomon (eds.): *Adhesion and Adhesives,* 2nd ed., **vol. 1,** Chap. 2, Elsevier, Amsterdam-London-New York 1965.

[41] H. Reinhard: "Dispersionen synthetischer Hochpolymerer," part II: Anwendungen, *Chemie, Physik und Technologie der Kunststoffe in Einzeldarstellungen,* vol. 14, Springer Verlag, Berlin-Heidelberg 1969.

[42] R. L. Patrick (ed.): *Adhesion and Adhesives,* **vol. 2: Materials,** Marcel Dekker, New York 1969.

[43] Lord Corp., DE 2 136 337, 1971 (Duston).

[44] L. Skeist (ed.): *Handbook of Adhesives,* Van Nostrand Reinhold, New York 1977.

[45] Ch. Gloxhuber, J. Galinke, *Chem. Ing. Tech.* **40** (1968) 485–486.

[46] V. K. Krieble, US 2 895 950, 1959.

[47] H. Lee, K. Neville: *Handbook of Epoxy Resins,* Chap. 21, McGraw Hill Book Comp. (1967)

[48] L. Jackson, *Adhes. Age* **6,** (1963) 34.

[49] R. J. Arnold, *Mod. Plast.* **41** (1964) 149–160.

[50] B. A Dombrow: *Polyurethanes,* Van Nostrand Reinhold, New York 1957.

[51] R. M. Carvey, *Adhes. Age* **24** (1968) 35–37.

[52] W. M. Ramsay: *Epoxy and Urethane Adhesives,* Palmerton Publ. Co., New York 1971.

[53] H. Lee, D. Stoffey, K. Neville: *New Linear Polymers,* McGraw Hill, New York 1967, p. 207.

[54] Henkel, DE 204 1126, 1970 (Klement, Budnowski, Lachmann).

[55] K. Katz, C. V. Cagle: *Adhesive Materials,* Forster Publ. Co, Long Beach, USA, 1971.

[56] W. H. Cobbs, *Adhes. Age* **22** (Dec. 1979) 31–36.

[57] W. Brockmann: "Grundlagen u. Stand der Metallklebtechnik," *VDI Taschenbücher T 22,* VDI-Verlag, Düsseldorf 1971.

[58] *"Testing of Adhesives",* TAPPI Monogr. Ser. 1974, no. 35.

[59] American Society for Testing and Materials, Philadelphia, PA, USA.

[60] Deutsches Institut f. Normung, Berlin, s.a. DIN-Normblattverzeichnis (annual) Beuth Verlag, Berlin-Köln.

[61] British Standards Institution, London WI, GB.

[62] International Standards Organisation Geneva 20, Switzerland.

[63] R. L. Patrick (ed.): *Treatise on Adhesion and Adhesives,* **vol. 2: Materials,** Marcel Dekker, New York 1969.

[64] F. Kollmann: *Furnier, Lagenhölzer und Tischlerplatten,* Springer Verlag, Berlin 1962.

[65] F. Kollmann, E. Kuenzi, A. Stamm: *Principles of Wood Science and Technology,* Springer Verlag, Berlin-New York 1975.

[66] H. Baumann: *Leime und Kontaktkleber,* Springer Verlag, Berlin 1967.

[67] R. Köhler, L. Skark: "Duroplastische Kunststoffe für die Holzverleimung und Holzveredlung," in *Kunststoffhandbuch,* vol. X, Hanser Verlag, München 1968.

[68] C. Schuller: "Handwerkliche und industrielle Verleimungstechnik im Vergleich," *Bau Möbelschreiner* **5** (1981) 33–36.

[69] "Kunststoffrohre für den Bau", *Kunstst. Bau* **16** (1981) no. 2, 87–93.

[70] *Kleben von thermoplastischen Kunststoffen,* DVS-Verlag, Düsseldorf 1993.

[71] *Klebeanleitung PVC Druckrohre, Verklebung mit THF-Klebestoff,* Kunststoffrohrverein e.V., Bonn 1985.

[72] R. Nowack, E. Barth: "Kunststoffrohrleitungen im Bereich der Verfahrenstechnik," *3 R international* **20** (1981) no. 6/7, 337 – 353.

[73] H. Fühn: "Kunststoffrohre aus PVC-hart in der Gasverteilung," *3 R international* **17** (1978) no. 10, 611 – 615.

[74] G. Poschet, I. Zöhren: "Chemikalienbeständigkeit von Klebverbindungen," *Schweißen + Schneiden* **31** (1979) no. 8, 332 – 335.

[75] W. Trautvetter: "Über chloriertes PVC," *Kunstst. Plast.***13** (1966) no. 2, 54 – 58.K. Binder, J. Fabing: Beständigkeit von PVC/PVC und PVC/Aluminiumverklebungen gegen verschiedenartige äußere Einflüsse, sterreichisches Kunststoffinstitut, Wien.Adhesives for Pipes and Tiltings of Unplasticized PVC, ISO 19E, 20E, 21E, International Standard Organisation.P. Benjamin: "Quality and Quality Control of Unplasticized Polyvinylchloride (uPVC) Pressure Pipes," *Plast. Rubber: Mater. Appl.5* (1980) no. 4, 151 – 160.

[76] S. Boström: *Kautschuk-Handbuch,* **vol. 4,** Berliner Union, Stuttgart 1961, p. 120.

[77] S. Buchan: *Rubber-to-Metal Bonding,* Crosby Lockwood, London 1959.

[78] G. Klement, *Gummi Asbest + Kunstst.* **24** (1971) 430 – 444.

[79] DeCrease: *Rubber World* **158** (1968) no. 1, 555 – 557.

[80] R. N. Özelli, H. Scheer: *Gummi Asbest + Kunstst.* **27** (1974) 612 – 618.

[81] C. V. Cagle: *Adhesive Bonding, Techniques and Applications,* McGraw-Hill, New York 1968.

[82] A. Matting: *Metallkleben,* Springer Verlag, Berlin - Heidelberg 1969.

[83] T. R. J. Schlieckelmann: *Metallkleben – Konstruktion und Fertigung in der Praxis,* Deutscher Verlag f. Schweißtechnik, Düsseldorf 1972.

[84] *Annual Book of ASTM Standards 1971,* part 16, Philadelphia, Pennsylvania.

[85] Military Specifications U.S. Army, Navy and Air Force Superintendent of Documents Washington 25, D.C.

[86] P. Wahl: *Handbuch Tapezieren-Kleben-Spannen,* Deutsche Verlagsanstalt, Stuttgart 1974

[87] Bundesausschuß Farb- und Sachwertschutz (ed.): *Technische Richtlinien für Maler- und Lackierarbeiten,* Nr. 16, Frankfurt 1976.

[88] P. Fickeisen, R. Füßl, R. Hummerich, I. Neumann, *Adhäsion* **42** (1998) no. 1/2, 24 – 29.

[89] D. Urban, E. Wistuba in D. Distler (ed.): *Wäßrige Polymerdispersionen,* Wiley-VCH, Weinheim 1999, pp. 150 – 154.

[90] Rohm & Haas, *Adhäsion* **42** (1998) no. 1/2, 25 – 29.

[91] *European Adhesives & Sealants,* **6** (1997) 28.

[92] G. Blunk, *KIB Kunstst. Bau* **9** (1968) 17 – 20.

[93] G. Tauber, *Seifen Öle Fette Wachse* **92** (1966) 899 – 908.

[94] M. Michel, *Tech. Mitt.* **65** (1972) 590 – 594.

[95] F. Mang, *Adhäsion* **23** (1979) no. 1, 4 – 9.

[96] R. Hinterwaldner, *Holzzentralbl.* **107** (1981) 180 – 181.

[97] R. Hinterwaldner, *Adhäsion* **21** (1977) no. 1, 13 – 20.

[98] R. Petri, *Boden Wand Decke* 1977, no. 11, 65 – 72.

[99] S. Schaaf, *Textilveredlung* **9** (1974) 14 – 25.

[100] S. Schaaf, H. Luthi, *Textilveredelung* **11** (1976) 348 – 351.

[101] D. V. Kamat, *Text. Chem. Color.* **5** (1973) 25 – 27.

[102] E. de Jong, *Adhäsion* **19** (1975) 317 – 324.

[103] S. Schaaf, *Am. Dyest. Rep.* **61** (1972) 31 – 40.D. E. Peermann, *Adhes. Age* **24** (1981) no. 12, 23 – 27.

[104] U. Maag, *Adhäsion* **20** (1976) 173 – 177, 206 – 209.

[105] W. Eichler, *Silikattechnik* **17** (1966) 227 – 228

[106] W. Moebes, A. Wende, *Plaste Kautsch.* **9** (1962) 232–236

[107] W. Dümig, F. Geyer, *Adhäsion* **22** (1978) no. 3, 70–74; **22** (1978) no. 4, 107–118; **22** (1978) no. 5, 145–147; **22** (1978) no. 8, 252–261; **22** (1978) no. 9, 288–291.

[108] R. Becker, *Plaste Kautsch.* **14** (1967) 96–98.

[109] F. Moser: "Bonding Glass," *Handbook of Adhesives,* Reinhold, New York 1962.

[110] *Coating* **2** (1969) 182.

[111] W. Brockmann, *Adhäsion* **22** (1978) 107–108.

[112] J. Schütte: *Der Luftschiffbau Schütte Lanz 1909– 1925,* R. Oldenbourg, München 1926.

[113] R. Houwink, G. Salomon: *Adhesion and Adhesives,* Elsevier, Amsterdam-London-New York 1965.

[114] E. Golf: "De Havilland Hornet," *Aircraft Prod.* **8** (1946) no. 4, 211.

[115] R. J. Schliekelmann: "Operational Experience with Adhesive Bonded Structures," in: Bonded Joints and Preparation for Bonding, *AGARD-Lecture Ser.* **1979,** no. 102, 1–1/1–30.

[116] R. J. Schliekelmann: *Metallkleben-Konstruktion und Fertigung,* DVS-Fachbuchreihe Schweißtechn., vol. 60, Deutscher Verlag f. Schweißtechnik, Düsseldorf 1972.

[117] E. Thrall: "Failures in Adhesively Bonded Structures," in: Bonded Joints and Preparation for Bonding, *AGARD-Lecture Ser.* **1979,** no. 102, 5–1/5 –89.

[118] J. E. Mc Carty: "Application of Advanced Structures and Materials to Transport Aircraft Fuselages," *15th Structures, Structural Dynamics and Materials Conference,* AIAA, New York 1974.

[119] A. Bethune: "Die Beständigkeit geklebter Aluminiumkonstruktionen," *Adhäsion* **20** (1982) no. 12, 347.

[120] W. Brockmann, O.-D. Hennemann, H. Kollek: "Surface Properties and Adhesion in Bonding Aluminium Alloys by Adhesives," *Int. J. Adhesion* **2** (1982) no. 1, 33/40.

[121] D. Schulz: "Airbus-Seitenruder in Faserverbundwerkstoffen," *K Kunststoffberater*1982, no. 4, 18/28.

[122] W. E. Quist, G. H. Narayanan, A. L. Wingert: "Aluminium-Lithium Alloys for Aircraft Structure, an Overview," 2. International Aluminium-Lithium Conf., Monterey, 12. 4.–14. 4. 1983.

[123] Skeist Laboratories Inc., Livingston New Jersey.

[124] Japan Adhesive Industry Assoc., Tokyo.

[125] Frost & Sullivan Inc., New York.

[126] Stat. Bundesamt Wiesbaden, FRG.

[127] Strategic Analysis, Reading, Pennsylvania, USA.

[128] Industrial Aids, London, GB.

[129] E. H. Schindel-Bidinelli, W. Gutherz: *Konstruktives Kleben,* VCH Verlagsgesellschaft, Weinheim 1998, pp. 1–4.
R. Houwink, G. Salomon: *Adhesion and Adhesives,* **vol. 1: Adhesives,** Elsevier, Amsterdam–London–New York 1965, pp. 1, 543–546.

[130] G. Habenicht: *Kleben,* 3rd ed., Springer, Berlin 1997, p. 295.

[131] W. Brockmann, *J. Adh.* **37** (1992) 173–179.

[132] EN 923, Adhesives—Terms and definitions, January 1998

[133] DIN EN 233, Feb. 1997; DIN 234, Feb. 1997; DIN 235, "Wandbekleidungen in Rollen," Oct. 1989.

[134] G. S. Haviland: *Machinery Adhesives for Locking, Retaining and Sealing,* Marcel Dekker, New York 1986.

[135] C. W. Boeder in S. R. Hartshorn (ed.): *Structural Adhesives Chemistry and Technology,* Chapter 5, Plenum, New York 1986, pp. 217–247.

[136] J. M. Rooney, B. M. Malofsky in I. Skeist (ed.): *Handbook of Adhesives,* 3rd. ed., Chapter 26, Van Nostrand Reinhold, New York 1990, pp. 451–462.

[137] R. D. Rich in A. Pizzi, K. Mittal (eds.): *Handbook of Adhesive Technology,* Chapter 29, Marcel Dekker, New York 1994, pp. 467–479.

[138] D. J. Stamper, *Brit. Polymer J.* **15** (1983) 34–39.

[139] Y. Okamoto, *J. Adhesion* **32** (1990) 227–235.

[140] Y. Okamoto, *J. Adhesion* **32** (1990) 237–244.

[141] T. Okamoto, H. Matsuda, *Nippon Setchaku Kyokaishi* **20** (1984) no. 10, 468–471.

[142] S. J. Hudak, F. J. Boerio, P. J. Clark, Y. Okamoto, *Surface and Interface Analysis* **15** (1990) 167–172.

[143] General Electric, US 2 628 178, 1950 (R. E. Burnett, B. W. Nordlander).

[144] American Sealants, US 2 895 950, 1957 (V. K. Krieble).

[145] Loctite, US 3 300 547, 1964, (J. W. Gorman, B. W. Nordlander).

[146] Loctite, US 3 425 988, 1965 (J. W. Gorman, A. S. Toback).

[147] Loctite, US 4 018 851, 1975, US 4 295 909, 1980, US 4 309 526, 1980 (L. J. Baccei).

[148] Loctite, US 3 435 012, 1965 (B. W. Nordlander).

[149] Princeton Chemical Research, US 3 428 614, US 3 451 980, 1966 (A. M. Brownstein).

[150] Princeton Chemical Research, US 3 595 969, 1969 (T. H. Shepherd, F. E. Gould).

[151] Eschem, US 4 209 604, 1976, US 4 431 787, 1981 (G. P. Werber).

[152] Three Bond, JP 44 007 541, 1969 (S. Kiyono, R. Ogawa).

[153] Three Bond, US 3 890 273, 1974 (T. Saito).

[154] Henkel, US 3 642 750, 1967 (B. Wegemund, J. Galinke).

[155] Rohm und Haas, US 4 234 711, 1978 (W. D. Emmons, V. J. Moser).

[156] Dow Corning, US 4 035 355, 1975 (R. H. Baney, O. W. Marko).

[157] Toshiba Silicone, EP 467 160, 1992 (F. Tetsuo, T. Masahiro, E. Isao).

[158] Loctite, US 5 605 999, 1995 (H.-K. Chu, R. D. Rich, I. D. Crossan, R. P. Cross).

[159] Loctite, US 5 635 546, 1996 (R. D. Rich, E. Maandi, P. M. Gontarz, H.-K. Chu).

[160] Toagosei, US 3 925 322, 1974 (K. Azuma et al.).

[161] Borden, US 3 795 641, 1971 (W. A. Lees, D. J. Bennet, J. R. Swire, P. Harding).

[162] American Sealants, US 3 041 322, 1959 (V. K. Krieble).

[163] American Sealants, US 3 046 262, 1960 (V. K. Krieble).

[164] Loctite, US 3 218 305, 1963 (V. K. Krieble).

[165] Loctite, US 4 180 640, 1977 (D. P. Melody, D. A. Doherty, J. F. O'Grady, R. D. Rich).

[166] Loctite, US 4 287 330, US 4 321 330, 1975 (R. D. Rich).

[167] Henkel, US 3 984 385, 1975 (W. Gruber, J. Galinke, J. Keil).

[168] Henkel, US 3 985 943, 1974 (W. Gruber, J. Galinke, J. Keil).

[169] W. R. Grace, US 4 429 063, 1982 (K. Reich).

[170] Loctite, US 4 513 127, 1984 (A. F. Jacobine).

[171] Loctite, US 4 622 348, 1984 (A. F. Jacobine, D. M. Glaser).

[172] American Sealants, US 3 043 820, 1960 (R. H. Krieble).

[173] Loctite, US 4 038 475, 1975 (E. Frauenglass, G. P. Werber).

[174] Loctite, US 3 625 875, 1968 (E. Frauenglass, W. E. Cass).

[175] Loctite, US 3 547 851, 1968 (E. Frauenglass).

[176] Borden, US 3 419 512, 1965 (W. A. Lees, J. R. Swire).

[177] Loctite, US 3 794 610, 1971 (A. G. Bachman).

[178] Borden, US 4 138 449, 1976 (T. R. Baldwin, D. J. Bennett, W. A. Lees).

[179] Loctite, US 3 988 299, 1975 (B. M. Malofsky).

[180] L. J. Baccei, B. M. Malofsky, *Polym. Sci. Technol.* **29** (1984) 589 – 601.
[181] Loctite, US 4 813 714, 1988 (C. B. Fairey, E. Frauenglass, L. W. Vincent).
[182] Loctite, U.S. Patent 3 591 438, 1968 (A. S. Toback, J. T. O'Conner).
[183] Loctite, US 3 616 040, 1968 (A. S. Toback).
[184] Loctite, US 3 625 930, 1967 (A. S. Toback, W. E. Cass).
[185] H. Onusseit, *Adhesives & Sealants Agenda Europe,* Campden Publishing, London 1997, p. 78.
[186] H. Onusseit, *Bindereport* (1997) no. 6, p. 373 – 381.
[187] H. Onusseit, *European Adhesives & Sealant* (1990) no. 3, p. 26 – 32.
[188] H. Onusseit, *Binderreport* (1996) no. 9, p. 465 – 473.
[189] P. Allbericci: "Aerospace Applications" in A. J. Kinlock (ed.): *Durability of Structural Adhesives,* Chap. 8, Applied Science Publishers, London 1983.
[190] E. W. Thrall, R. W. Shannon (eds.): *Adhesive Bonding of Aluminum Alloys,* Marcel Dekker, New York – Basel 1985.
[191] W. Brockmann, O.-D. Hennemann, H. Kollek, C. Matz, "Adhesion in Bonded Aluminum Joints for Aircraft Construction", *Int. J. Adhes. Adhes.* **6** (1986) 115 – 143.
[192] EN 2243 – 2, Aerospace series, Structural Adhesives, Test Methods, Peel Metal – Metal, AECMA, Bruxelles.
[193] EN 2243 – 6, Aerospace series, Structural Adhesives, Test Methods, Determination of Shear Stress and Shear Strain, AECMA, Bruxelles.
[194] US 3 959 091, 1971 (Y. Moji, J. A. Marceau).
[195] C. W. Matz, "Optimization of the Durability of Structural Titanium Adhesive Joints", *Int. J. Adhes. Adhes.* **8** (1988) 17 – 24.
[196] Y. Ika: "Tissue Adhesives," in C. C. Chu, J. A. von Fraunhofer, H. P. Greisler (eds.): *Wound Closure Biomaterials and Devices,* CRC Press, Boca Raton, FL 1996, pp. 317 – 346.
[197] D. Sierra, R. Slatz (eds.): *Surgical Adhesives and Sealants,* Technomic Publishing Company, Lancaster, PA 1996.
[198] G. Lewis, *J. Biomed Mater. Res. (Appl. Biomater.)* **38** (1997) 155 – 182.
[199] Loctite Corporation (ed.): *Loctite Worldwide Design Handbook,* 1996/97 ed., 1995.
[200] J. H. Bickford: *An Introduction to the Design and Behavior of Bolted Joints,* 2nd ed., Marcel Dekker, New York 1990, p. 462 – 463.
[201] C. L. Brett, *Int. J. Adhes. Adhes.* **2** (1982) no. 1, 19 – 24.
[202] E. Maandi, R. D. Rich, *Scientific Computing and Automation* (1994) Aug., 23 – 27.
[203] US 3965, 1845 (H. Day, W. H. Shecut).
[204] C. W. Bemmels: "Pressure-Sensitive Tapes and Labels" in I. Skeist (ed.): *Handbook of Adhesives,* 2nd ed., Van Nostrand Reinhold, New York 1977, p. 724.
[205] F. H. Wetzel, *Rubber Age* **82** (1957) 291 – 295.
[206] C. W. Hock, A. N. Abbott, *Rubber Age* **82** (1957) 471 – 475.
[207] K. Kamagata, H. Kosaka, K. Hino, M. Toyama, *J. Appl. Polymer Sci.* **15** (1971) 483 – 500.
[208] M. Sherriff, R. W. Knibbs, P. G. Langley, *J. Appl. Polymer Sci.* **17** (1973) 3423 – 3438.
[209] "Pressure Sensitive Adhesives" in *Encyclopedia of Polymer Science and Engineering,* **vol. 13,** John Wiley & Sons, New York 1988.
[210] J. D. Ferry: *Viscoelastic Properties of Polymers,* 2nd ed., John Wiley & Sons, New York 1961.
[211] D. Satas (ed.): *Handbook of Pressure Sensitive Technology,* 2nd ed., Van Nostrand Reinhold, New York 1982.
[212] W. A. Zisman in: "Contact Angle, Wettability and Adhesion," *ACS Advances in Chemistry Series* (1964) no. 1.
[213] M. Toyama, T. Ito, H. Morigushi, *J. Appl. Polym. Sci.* **14** (1970) 2039 – 2048.

[214] M. Toyama, T. Ito, H. Morigushi, *J. Appl. Polym. Sci.* **14** (1970) 2095 – 2303.
[215] M. Toyama et al., *J. Appl. Polymer Sci.* **17** (1973) 3495 – 3502.
[216] A. Zosel, *Colloid Polym. Sci.* **263** (1985) 541 – 553.
[217] D. W. Aubrey, M. Sherriff, *J. Polym. Sci. Polym. Chem. Ed.* **18** (1980) 2597 – 2608.
[218] D. W. Aubrey: "Viscoelastic Basis of Peel Adhesion" in K. W. Allen (ed.): *Adhesion,* **vol. 3,** Applied Science Publishers Ltd., London 1978, p. 191 – 205.
[219] L. H. Sperling: *An Introduction to Physical Polymer Science,* John Wiley & Sons, New York 1983.
[220] *Test Methods for Pressure Sensitive Adhesive Tapes,* 11th ed., Pressure Sensitive Tape Council, 401 North Michigan Avenue, Chicago, IL 60611-4267 USA.
[221] ASTM American Society for Testing & Materials, 100 Barr Harbor Drive, West Conshohocken, PA 19428 USA.
[222] Association des Fabricants Europeens de Ruban Auto-adhesifs (A.F.E.R.A.), 60 rue Auber, 94408 Vitry sur Seine CEDEX, France, Manual 1983.
[223] European Committee for Standardization, 36, rue de Stassart, B-1050 Bruxelles, Belgium.
[224] T. B. Jensen, European Tape & Label Conference, Antwerp, April 1997, Belgium.
[225] Shell Chemicals, KRATON Elastomers — Test Methods for PSAs, July 1995.
[226] Exxon Chemical, Manual Number EC - AIBU/Techn-QS-04/02, Procedure TAP 5 (30.10.1997).
[227] W. Hofman: *Rubber Technology Handbook,* Hanser Publishers, Munich 1996.
[228] 3M Company, US 2 956 904, (J. O. Hendricks).
[229] 3M Company, WO 94/11175 and WO 95/25774, (T. D. Bredahl et al.).
[230] M. F. Tse, *J. Adhes. Sci. Technol.* **3** (1989) no. 7, 551 – 570.
[231] G. R. Himes, L. H. Oliveri, *Adhes. Age* 1998, Feb., 41 – 44.
[232] J. R. Erickson, Adhesives Age 1986, April.
[233] M. DuPont, N. De Keyzer, "UV-Curing of Styrenic Block Copolymer Based Pressure Sensitive Adhesives," RadTech Europe, Maastricht 1995, p. 174 ff.
[234] R. Milker, *Z. Czech Adhaesion* **3** (1985) 29 – 32.
[235] W. Karmann, S. Zöllner: "Radiation Curing of Pressure Sensitive Adhesives: An Overview," RadTech Europe, Maastricht 1995, p. 155 ff.
[236] 3M Comp., US 4 181 752, (J. A. Martens et al.).
[237] G. Auchter, J. Barwich, G. Rehner, H. Jager, *Adhesives Age* 1994, July, 20 – 25.
[238] A. B. Kummer: "Vernetzung von Acrylat-Schmelzhaftklebmassen," 23th Munich Adhesive and Finishing Seminar, 1998.
[239] R. B. Fox in: *Handbook of Chemistry and Physics,* 76th ed., CRC New York, London 1996.
[240] D. F. Merill: "Silicone Pressure Sensitive Adhesives" in D. Satas (ed.): *Handbook of Pressure Sensitive Technology,* 1st ed., Van Nostrand Reinhold, New York 1982.
[241] S. R. Kerr: "Recent Advances in Silicone PSA Technology" PSTC Tech XVII Seminar Proceedings Hyatt Regency, Schaumburg IL, 4th – 6th May 1994, p. 187 – 201.
[242] ASTM D 2398, Ring & ball softening temperatures.
[243] ASTM E 28-27 "Standard Test Method for Softening Point of Resins by Ring and Ball Apparatus," pub. Sept. 1997.
[244] Hercules Inc.: "The Hercules Spectrum of Hydrocarbon Resins," bulletin 900-204, Wilmington, DE.
[245] 3M, EP 0 202 938, 1985 (G. F. Vesley, A. H. Paulson, E. C. Barber).
[246] 3M, EP 0 257 984, (Darvell, Konsti, Klingen, Kreckel).
[247] J. Murphy: *Additives for Plastics Handbook,* Elsevier Science, Amsterdam 1996, p. 116 – 118.
[248] DuPont de Nemours, US 2 273 040, (R. K. Iler).
[249] Beiersdorf, DE 2 845 541, 1978 (P. Gleichenhagen et al.).

[250] 3M, US 3 318 852, (G. M. Dixon).
[251] Permacel, US 4 513 059, (W. C. Dabroski).
[252] J. D. Jones, Y. A. Peters: "Silicone Release Coatings" in [211] p. 601 – 626.
[253] Th. Ebbrecht, P. Lersch, D. Wewers: "Radiation Curing of Silicone Acrylate Systems," RadTech Europe, Mediterraneo 1993, p. 322 – 328.
[254] R. P. Eckberg, B. D. Shepherd: "Radiation Curable Silicone Coatings: Controlled Release Design," RadTech North America, Orlando 1994, p. 507 – 514.
[255] E. D. Cohen, E. B. Gutoff (eds.): *Modern Coating and Drying Technology,* VCH Verlagsgesellschaft, Weinheim 1992.
[256] R. Zimmet: "Continuous Production of Solventless Adhesives," European Tape & Label Conference, April 1997, Antwerp, Belgium.
[257] P. Franz: "Kontinuierliche Aufbereitung ... Schmelzkleber zur off-line und in-line Beschichtung von Substraten," 22. Münchener Klebstoff- und Veredlungs-Seminar, 1997, p. 6 – 11.
[258] C. Rauwendaal: *Polymer Extrusion,* 2nd ed., Hanser Publishers, Munich 1990.
[259] P. M. Schweizer in [255], p. 23 – 61.
[260] F. Shepherd: *Modern Coating Technology Systems,* Emap McLaren Ltd., Barnet 1995, p. 13 – 22.
[261] P. Herzog: "Neue Trends auf dem Sektor der Hotmelt-Düsenbeschichtung," 22. Münchener Klebstoff- und Veredlungs-Seminar, 1997, p. 1 – 5.
[262] Nabert-Schön: *Sicherheitstechnische Kennzahlen brennbarer Gase und Dämpfe,* Deutscher Eich Verlag, Braunschweig 1978 (corr. 1.7.1990).
[263] W. Karmann, S. Zöllner: "Radiation Curing of Pressure Sensitive Adhesives: An Overview," RadTech Europe, Maastricht 1995.
[264] W. Karmann, *J. of Irradation Tech.* **1** (1983) no. 4, 305 – 323.
[265] U. Köhler, G. Guse, RadTech Europe, Mediterraneo 1993.
[266] G. G. Skelhorne: "Electron Beam Processing" in *Chemistry & Technology of UV & EB Formulations for Coatings, Inks & Paints,* **vol. 1,** SITA Technology, London 1991, pp. 103 – 157.
[267] R. E. Knight: "UV Curing Equipment and Applications" in Chemistry and Technology of UV and EB Formulations for Coatings, Inks and Paints, vol. 1, SITA Technology, London 1991, pp. 161 – 241.
[268] S. Röber: "Packaging Tapes" in D. Satas (ed.): *Handbook of Pressure Sensitive Adhesive Technology,* 3rd ed., to be published.
[269] *Ullmann's,* 5th ed., **A26,** p. 97 – 100.
[270] DuPont, Kaladex, Technical Data Sheet KX TD 003.
[271] UL Underwriters Laboratories Inc., 333 Pfingsten Road, Northbrook IL, 60062, USA.
[272] VDE Verband Deutscher Elektrotechniker e.V. VDE-Verlag GmbH, Bismarckstr. 33, D-10625 Berlin, Germany.
[273] DIN Deutsches Institut für Normung e.V. Beuth Verlag GmbH, Burggrafenstr. 6, D-10787 Berlin, Germany.
[274] IEC International Electrotechnical Commition, 3 Rue de Varembe, CH-1211 Geneva, Switzerland.
[275] K. Külper: "Electrical Tapes" in D. Satas (ed.): *Handbook of Pressure Sensitive Adhesive Technology,* 3rd ed., in press.
[276] Beiersdorf, EP 58 382, 1982 (Gleichenhagen et al.).
[277] Johnson & Johnson, US 4 024 312, 1976 (R. Korpman).
[278] Beiersdorf, EP 0 578 979, 1992 (B. Lühman).
[279] 3M, US 4 223 067, 1977 (D. Levens).
[280] Beiersdorf, DE 19 614 620, 1996 (U. Schümann).

[281] G. Bennet, J. Klingen, *Adhes. Age* 1996, Sept.

[282] P. G. Unna: "Weißes Kautschukheftpflaster" *Monatsh. Prakt. Dermatol. (Hamburg)* **29** (1899) 511 – 514.

[283] 3M, US 2 884 126, 1958 (E. W. Ulrich).

[284] E. Riedel, W. Triebsch, K. M. Sedlarik: *Verbandstoff-Fibel,* 5th ed., Wissenschaftliche Verlagsgesellschaft, Stuttgart 1995.

[285] I. Webster, *J. Adhes.* **17** (1997) no. 1, 69 – 73.

[286] S. A. Huie, P. F. Schmit, J. Swenson Warren, *Adhes. Age* **28** (1985) no. 6., 30 – 35.

[287] D. Satas, A. M. Satas, in [10], p. 637.

Index